Table of Contents

Introduction 9

To The Student 11

Chapter 1: Modeling Preliminaries 13

    1.1: How Models are Created 15

        1.1.1: Modeling Cycle by Example: The Literary Digest Case 15

        1.1.2: Modeling Cycle by Example: Weather Forecasting 17

        1.1.3: Cautions on Modeling 18

    1.2: Using Equations 21

        1.2.1: Evaluating Equations 21

        1.2.2: Issues in Evaluating Equations 24

    1.3: Dimensional Analysis 29

        1.3.1: Adding and Subtracting Units 29

        1.3.2: Multiplying Units 31

        1.3.3: Dividing Units 33

    1.4: Errors 39

        1.4.1: Measures of Error 39

        1.4.2: Types of Errors 43

        1.4.3: Error Tables 46

    1.5: Application: Some Common Models 54

        1.5.1: Measuring Space: Area and Perimeter 54

        1.5.2: Distance-Rate-Time 58

Chapter 1 Summary 64

Chapter 2: Data Collection 67

    2.1: Common Terms in Data Collection 69

©Brian Gillispie, 2016.

    2.1.1: Types of Variables 69

    2.1.2: Levels of Measurement 71

    2.1.3: Types of Models 73

  2.2: Independent and Dependent Variables 77

    2.2.1: Definitions 77

    2.2.2: Selecting Variables from Problems 80

  2.3: Numerical Scales 85

    2.3.1: Using a Scaled Model 85

    2.3.2: Data Point Scaling 87

  2.4: Handling Categorical Data 93

    2.4.1: Converting Categorical Data to Quantitative Data 93

    2.4.2: Some Issues 96

  2.5: Domain and Range 100

    2.5.1: Domain 100

    2.5.2: Range 104

    2.5.3: Domain and Range of Known Functions 104

  2.6: Application: Ethics and Data Collection 113

    2.6.1: Data Collecting Issues 113

    2.6.2: Data Reporting Issues 115

Chapter 2 Summary 118

Chapter 3: Measuring Change 121

  3.1: Measuring Change 123

    3.1.1: Total Change 123

    3.1.2: Average Rate of Change 125

    3.1.3: Arithmetic Change 128

    3.1.4: Geometric Change 129

©Brian Gillispie, 2016.

3.2: Interpretation of Change 134

3.2.1: Interpreting Average Rate of Change 134

3.2.2: Interpreting Geometric Change 135

3.2.3: Putting it All Together 136

3.3: Using Change: Single Point Modeling 142

3.3.1: Using the Average Rate of Change 142

3.3.2: Using the Geometric Change 145

3.3.3: Combined Change 148

3.4: Using Change: Two Point Modeling 153

3.4.1: Using the Average Rate of Change 153

3.4.2: Using the Geometric Change 157

3.4.3: Putting it all Together 159

3.5: Change Tables 164

3.6: Using Change: Three Point Modeling, Sequential Points 170

3.6.1: Change of the Change: Idea and Definition 170

3.6.2: Using the Change of the Change 173

Chapter 3 Summary 182

Chapter 4: Discrete Modeling with Sequences 185

4.1: Sequences 187

4.1.1: What is a Sequence? 187

4.1.2: Sequence Notation 188

4.2: Recursive Sequences 192

4.2.1: Recursive Sequence Representation 192

4.2.2: Calculating terms in a Recursive Sequence 193

4.2.3: Excel Representation of Recursive Sequences 196

4.3: Creating Recursive Sequences 201

©Brian Gillispie, 2016.

- 4.3.1: Sequences from Two Points ... 201
- 4.3.2: Sequences from Rates of Change ... 204
- 4.4: Long Term Behavior of Recursive Sequences ... 209
  - 4.4.1: Fixed Points of Recursive Sequences ... 209
  - 4.4.2: Stability of Fixed Points ... 213
  - 4.4.3: Applications of Stability and Fixed Points ... 217
- 4.5: Interconnected Sequences ... 224
- Chapter 4 Summary ... 234

Chapter 5: Linear Models ... 235

- 5.1: Review of Lines ... 237
  - 5.1.1: Slope ... 237
  - 5.1.2: The Y Intercept ... 240
  - 5.1.3: Definition of a Line ... 241
  - 5.1.4: Point-Slope Form of a Line ... 243
- 5.2: Modeling with Slopes and Intercepts ... 248
  - 5.2.1: Interpreting Slope ... 248
  - 5.2.2: Interpreting Y - Intercepts ... 249
  - 5.2.3: Putting it all Together ... 250
- 5.3: Linear Models from Two Data Points ... 256
- 5.4: Applications of Linear Models ... 263
  - 5.4.1: Cost and Revenue Models ... 263
  - 5.4.2: Supply and Demand Models ... 269

Chapter 5 Summary ... 279

Chapter 6: Exponential Models ... 281

- 6.1: Review of Exponents and Logarithms ... 283

- 6.1.1: Exponents 283
- 6.1.2: Logarithms and Solving Exponential Equations 285
- 6.2: Creating an Exponential Model 291
- 6.3: Shifting the Time Interval 301
- 6.4: Half Life and Doubling Time Models 309
- 6.5: Exponential Models from Two Data Points 319
- 6.6: Application: Examples of Some Exponential Models 327
  - 6.6.1: Continuous Growth Model 327
  - 6.6.2: The Logistic Model 331
  - 6.6.3: Newton's Law of Cooling 336

Chapter 6 Summary 342

Chapter 7: Other Continuous Models 345

- 7.1: Some Quadratic Modeling 347
  - 7.1.1: The Height Equation 347
  - 7.1.2: The Zeros of a Quadratic 348
  - 7.1.3: Maximums or Minimums of Quadratic Models 354
- 7.2: Polynomial Interpolation 359
  - Appendix to Section 7.2: Calculator Directions 367
- 7.3: Proportionality 371
  - 7.3.1: Direct Proportionality 371
  - 7.3.2: Inverse Proportionality 374
- 7.4: Trigonometric Modeling 383
  - 7.4.1: Common Terms 383
  - 7.4.2: Creating Trigonometric Models 388
- 7.5: Log Scaling 397

©Brian Gillispie, 2016.

    7.5.1: The Idea 397

    7.5.2: Creating a Log Scaled Model 401

Chapter 7 Summary 407

Chapter 8: Model Selection 411

  8.1: Review of Mean and Standard Deviation 413

    8.1.1: Mean 413

    8.1.2: Standard Deviation 416

    8.1.3: Using Technology 422

  8.2: Method of Change Comparison 428

  8.3: Method of Least Error 438

  8.4: Method of Lowest Squared Error 447

  8.5: Modeling Categorical Data 457

Chapter 8 Summary 468

Chapter 9: Other Topics 471

  9.1: Zero Finding 473

    9.1.1: Bisection Method 473

    9.1.2: Secant Method 478

  9.2: Domain Restricting 482

  9.3: Piecewise Models 489

    9.3.1: Reading Piecewise Models 489

    9.3.2: Creating Piecewise Models 491

  9.4: Parametric Models 501

    9.4.1: Definition of Parametric Models 501

    9.4.2: A Common Parametric Model: Modeling Height and Distance 504

Chapter 9 Summary 511

©Brian Gillispie, 2016.

Appendix A: Challenge Problems 513

Appendix B: Answers to Selected Exercises 519

Index 545

References 549

©Brian Gillispie, 2016.

©Brian Gillispie, 2016.

Introduction

This book was written in an attempt to introduce the world of mathematical modeling to students who have not had calculus or beyond. Most of the mathematical modeling books currently on the market assume at least Calculus I in their reader's background, and a high percentage of the books assume the reader has also had Calculus II, Linear Algebra, and Differential Equations. So in an attempt to bridge that gap and introduce mathematical modeling to the rest of the world who have not had differential equations, I am writing this book.

This first edition is used to teach a course called Mathematical Modeling at Mount Mercy University in Cedar Rapids, Iowa. The students in this class have not taken Calculus I, but have taken at least two years of high school mathematics. This book will assume that the reader is familiar with and understands material commonly taught in a high school algebra class. This includes but is not limited to how to solve linear equations, order of operations, how to convert to and from different units, how to solve quadratics, and what an exponent is, just to name a few.

It is recommended to read the Chapters in order, as material in Chapters assume that the reader is comfortable with the definitions they have encountered so far in the book. The Prerequisites for each Chapter are as follows:

Chapter 1: None

Chapter 2: Chapter 1

Chapter 3: Chapters 1 - 2

Chapter 4: None for Sections 4.1, 4.2, Chapter 3 for 4.3 - 4.5.

Chapter 5: Chapters 1 - 2.

Chapter 6: Chapters 1 - 3.

Chapter 7: Chapters 1 and 2 for 7.1 - 7.4, Chapter 5 for 7.5.

Chapter 8: Chapters 1, 2, 3, 5, 6, 7. Sections 8.2 - 8.4 require Sections 5.3 and 6.5.

Chapter 9: Chapters 3 - 7 for 9.1, Chapters 1, 2, 3, 5 - 7 for 9.2, 9.3, 9.4 and 9.5.

©Brian Gillispie, 2016.

Note that if you are short on time, you can safely skip the following Sections, as the topics covered here are independent and rarely mentioned outside their Sections: 1.5, 2.6, 3.6 (unless covering 8.2), 4.4, 5.4, 6.5 (unless covering Chapter 8 and 9), 7.4, 7.5, 8.2, 9.1, 9.2.

A suggested flow for this book for a one semester three credit hour course that starts at the beginning of the book is as follows:

Chapter 1 -> Chapter 2 -> Chapter 3 -> Chapter 4 -> Chapter 5 -> Chapter 6 -> 7.1 -> 7.3 -> 9.4.

Be advised that Chapters 8 and 9 are thick, and will probably take multiple days to cover well. I advise the teacher to cover the topics from those Chapters that are of interest to them and their class.

If you have students who have seen the material in Chapters 1 and 2 before[1], you can safely start at Chapter 3. This should allow you to cover Chapters 3 - 8 in one semester, with any remaining time being used to cover Chapter 9.

A four credit course should be able to cover Chapters 1 - 8 in one term, with any extra time being used to cover Chapter 9.

A five credit course should be able to cover the entire book in one term, with some time left over to review concepts as needed.

---

[1] For example, if Precalculus is a prerequisite for your class, then you can probably skip Chapters 1 - 2, or lightly skim them.

©Brian Gillispie, 2016.

To The Student

It is my hope that you will find this book followable and it will help you to learn and appreciate mathematical models. Nevertheless there are a few things I do need to point out to you before proceeding.

First of all, some of the Exercises do build on previous Exercises in the Chapter. It is not recommended to skip Sections within a Chapter for that reason, except as noted in the introduction.

Secondly, you will occasionally see a * after the number of an exercise. What the * means is the exercise requires you to do outside research in order to solve the problem.

Thirdly, be advised that as we advance in Chapters some steps will be omitted, as we assume that by then you are able to perform the steps yourself. You will see this heavily in Chapters 8 and 9, as by then we assume you can work the problems in Chapters 1 - 7.

©Brian Gillispie, 2016.

©Brian Gillispie, 2016.

Chapter 1: Modeling Preliminaries

In our daily lives, we will often be confronted with various items that we wish to better understand. In an attempt to explain this phenomenon, we will sometimes use what is known as a mathematical model. By using a mathematical model, we can then better understand what is happening around us, as well as why it is happening.

In this Chapter we will start out by describing the process by which models are created. From there, we will cover how to evaluate the equations that arise from mathematical models, how to handle values that are in different units, and how error is calculated. Finally we will conclude by demonstrating a few well known models used for computing area and perimeter, as well as the distance-rate-time model.

©Brian Gillispie, 2016.

©Brian Gillispie, 2016.

1.1: How Models are Created

Mathematical models are created in an attempt to better understand the world around us. To create a mathematical model, first we have to collect the necessary data. However, not all data collected is useful, so we simplify the collected data appropriately. From there, we decide which mathematical formula (or formulas) fits the data we have, and create a model based on that. Next, the created model is run and results are found. These results are then analyzed to see what they tell us about the world around us. Then, the result of the model is compared to the real world result to check the model for accuracy. Based on the results of this step, the model may need to be adjusted, a different model may be needed, or more data may be needed. In all cases, we return to the beginning, update the model, and start over.

In this Section we will look at the modeling cycle, via two examples. We have not covered the mathematics behind how a model is created yet, so we will skip over that step. Later on, consider returning to this Section to see if you can create the same model we did in these Examples.

1.1.1: Modeling Cycle by Example: The Literary Digest Case

The first Example of the modeling cycle will be a classical historical example. Back in the year 1936, there was a presidential election, with the voters able to choose between FDR, who was the current president, and Alf Landon, his challenger [1] [2]. One magazine, the Literary Digest [3], wished to predict who was going to be voted for president in the upcoming presidential election.

In order to make this prediction, the magazine had to collect the necessary data. Now, there is no practical way the magazine could survey every single American and learn who they were going to vote for, so instead, the magazine chose to take a sample of ten million Americans. This sample of ten million Americans was collected from the database of telephone and automobile registrations on file at that time in 1936.

Once the magazine decided how the data was to be collected, they mailed out surveys to those ten million Americans selected for the survey. The magazine received back 2.5 million responses (a huge number even today for a survey), and in those surveys a high percentage of those who responded said they were going to vote for Alf Landon.

Based on the results of their survey, the Literary Digest analyzed the results and concluded that if high of a percentage of respondents in our survey are going to vote for Alf Landon, there is a very good chance that a high percentage of all voters are going to vote for Alf Landon. So, the Literary Digest interpreted the results of their survey, and concluded that Alf Landon will win the 1936 presidential election. Therefore, the magazine published a study saying that Alf Landon will win the election.

Once the study was published, all the magazine could do was wait for the real world results. Finally Election Day came, and FDR was reelected [1] [2]. Somehow, the magazine survey was wrong! The only thing to do at this point in time was return to the model and see where did the survey go wrong? How could a study of 2.5 million Americans not be an accurate representation of the real world? It turns out that the Literary Digest made a fatal choice in their choice of sample. Recall how they chose those who they mailed surveys to by the current telephone and automobile registrations? It turns out that in 1936, America was deep in the Great Depression, and only the rich had telephone and automobiles at that point in time! Therefore, by accident, the Literary Digest surveyed only the rich Americans, and their results were not representative of what the average American thought [3].

While the Literary Digest model was flawed, it still did serve a purpose, as it brought attention to how samples are conducted for models. Also, it illustrates that a huge sample size does not guarantee an accurate model, especially because in the same year Gallop polled only 50,000 Americans and had a more accurate result. To this day most studies are conducted via Gallop polls, and are fairly accurate despite a small sample size. For example, the Gallop student poll typically surveys less than one million people [3] [4], yet their results are considered very accurate and representative of how the average person feels about things.

So, while this first Example of the modeling cycle did result in a failed model, do not conclude that a model that returns a bad result is useless. Sometimes you can still learn important things from the model, despite the error. If nothing else, you learn what model to not use for the future.

©Brian Gillispie, 2016.

1.1.2: Modeling Cycle by Example: Weather Forecasting

Weather forecasting is a very complicated problem, as anyone who has been caught in a rainstorm when there is a zero percent chance of rain forecast for the day can tell you. However, we tend to do some very basic weather forecasting ourselves, very easily. Here is an Example.

In this case, pretend that you live in Minneapolis, Minnesota, and wish to know whether or not it will rain in 3 hours. Below are the weather reports from some of the surrounding cities:

City	Distance from	Wind	Weather
St. Cloud	66	22 N	71, Sunny
Mankato	81	25 NE	72, Thunderstorms
Brookings, SD	214	22 E	65, Rain
Portland, OR	1591	15 SE	52, Overcast

Obviously, some of these are not relevant to the problem at hand. For instance, the fact that it is raining in Portland, which is 1591 miles away, is not relevant to whether or not it is going to rain in 3 hours from now in Minneapolis, MN. Similarly, the weather in Brookings is unlikely to matter in this decision. Therefore, the data is simplified to consider only the weather from St. Cloud and Mankato, both nearby cities to the west of Minneapolis.

Now that the data is simplified, the mathematical model is created. Our model is going to be simple here. All we are going to do is take the distance and divide it by the wind speed. If that number is near 3, and if the wind is blowing in our direction, then chances are that weather will be our weather 3 hours from now.

With that decision made, we apply the model. First, let's consider the data we have from St. Cloud. St. Cloud is 66 miles away, and the wind is blowing to the north. 66/22 is 3 hours, so in 3 hours, our model says that the weather in St. Cloud will have moved 66 miles north of St. Cloud. However, a quick look at a map reveals that Minneapolis is not 66 miles north of St. Cloud, so we conclude that it will probably not be 71 and Sunny in 3 hours for us.

Next, let's apply this model to the data from Mankato. Mankato is 83 miles away, and the wind is blowing to the northeast. 83/25 is 3.32 hours. Also, the wind is blowing NE, and a quick glance at a map reveals that Minneapolis is NE of Mankato. Therefore, in 3.32 hours,

©Brian Gillispie, 2016.

Mankato's weather will be in Minneapolis, based on this model. Since Mankato is reporting thunderstorms, this means that most likely you will have thunderstorms in Minneapolis in about 3.32 hours.

Based on these results, and since you plan to leave in 3 hours, you would plan for thunderstorms. 3.32 hours is about 3 hours and 15 minutes, which means that by the time you leave it will either be storming or just about to. Better grab your rain coat.

Then, all you could do with this model is wait for 3 hours to pass, and see the results. In this case, thunderstorms do indeed arrive about 3 hours and 15 minutes, and you are glad you took your raincoat when you left. In this case, the model worked correctly, and there is no further updating to do to the model, so you file it away to use the next time.

Now, even though the model worked correctly this time, it might be the case that the next time you use this model it is incorrect. Maybe the storm suddenly heads north and misses you? In that case, you would note why the storm missed, and see if there is a way to update the model to account for those kinds of circumstances.

1.1.3: Cautions on Modeling

Before we conclude this Section, we wish to state a few cautions that a modeler needs to keep in mind when creating and using mathematical models.

Firstly, when creating a mathematical model, do not assume that it is useless because it has some error in the model. Most mathematical models have some sort of error. In fact, it is impossible to avoid all error in models, even with computers. Just because a model has an error or gives an unrealistic result at one point, do not assume that the model is completely useless. Sometimes the model is still useful for all points but that one. Sometimes the model still tells us some truth about how things operate, deep at the micro level (see [5] for an Example of this). In both cases, a full analysis of why the error is occurring is warranted, and can often still teach us a lot about how the world operates.

Secondly, know that mathematical modeling is an art. There is a lot of creativity involved. In modeling problems, no one is going to tell you what formula or procedure to use. Instead, you will have to experiment. See what happens if you try a linear model. If that does not work, maybe try an exponential model. See what happens, and modify your approach as needed.

©Brian Gillispie, 2016.

In this book, early on it will be pretty obvious what formula and approach to use. However, as we proceed through the various Chapters, which formula to use will be harder and harder to determine. My hope is by writing this book in this way, that you can build up the skills and creativity needed to become a successful mathematical modeler.

Still, do not despair if your models are wrong at first. Part of learning is making mistakes, and then fixing them and moving on. Edison did not create the light bulb correctly on his first try. The Wright brothers did not get a plane to fly on their first try. The US space program, in an attempt to land on the moon, had some major mishaps on the way as they learned how to launch a shuttle and have humans survive the attempt. The point is, failure is a part of learning something new. Learn from the mistakes, see what you did wrong, and then try to fix it next time around.

1.1 Exercises

1. For the model listed in Section 1.1.1, list the problem they were trying to solve, how the data was collected, the interpretation of the data, and what happened when the data was compared to real world data.

2. For the model listed in Section 1.1.2, list the problem we were trying to solve, the simplifying assumptions, the calculation applied to the data, the interpretation of the result, and what happened when the data was compared to real world data.

3. In the examples in this Section some of the steps appear to be missing. Can you think of an instance when a simplifying assumption would not be used on the data collected?

4. In the model in Section 1.1.1, no mathematical model was created or listed. Can you think of why this is?

5. No model works perfectly when compared against real world data. Therefore, how would you decide whether a model was accurate enough to use reliably and make real world predictions from? Be specific!

6. While models should always be tested against real world data, sometimes this data is unavailable or impossible to obtain, like for instance in a model of a nuclear explosion or a model of disease outbreaks. What ways can you think of to still test your model (legally and ethically) to make sure it is still a valid representation of the situation being modeled?

©Brian Gillispie, 2016.

1.2: Using Equations

Most mathematical models involve the use of equations. In this Section we will discuss equations and how to evaluate equations.

1.2.1: Evaluating Equations

In general, an equation is an expression of the form something is equal to something else, and this statement is true for various numerical values. Evaluating the equation means to plug in the given number or numbers and solve for the variable which is left over. When you evaluate the equation though, be careful to only plug in the number given, and do not lose any signs or operations on the way. Let's demonstrate now with a few Examples.

Example 1.2.1: Evaluate the equation $y = 2x + 5$ at $x = 3$

Solution to 1.2.1: The statement above means that wherever I see an x, I need to replace it with 3. Once I do that, I end up with:

$$y = 2(3) + 5 = 11$$

So, when x is 3, y is 11.

Example 1.2.2: Evaluate the equation $y = -16x^2 + 64x + 16$ at $x = 2$

Solution to 1.2.2: This statement means that wherever I see an x, I replace it with a 2. Be careful this time, as one of the terms in the equation has an x^2 in it, which means that when I replace x with 2, I do not lose the little 2 above it, as that must be applied to the value I plug in. Therefore, this means I end up with the following:

$$y = -16(2)^2 + 64(2) + 16$$

©Brian Gillispie, 2016.

This simplifies to:

$$y = -16(4) + 64(2) + 16 = 80$$

So, y is 80 when x is 2.

Example 1.2.3: Evaluate the equation $P = 2L + 2W$ for $L = 50$ and $W = 25$

Solution to 1.2.3: This statement means that wherever there is an L in the equation, replace it with 50, and wherever there is an W in the equation, replace it with 25. Doing that gives us the following:

$$P = 2 * 50 + 2 * 25 = 100 + 50 = 150$$

So when W is 25 and L is 50, P is 150.

Example 1.2.4: Evaluate the equation $y = 100 * 2^{x/5}$ for $x = 10$

Solution to 1.2.4: While this one looks different than the others, the idea is the same, wherever there is an x, replace it with 10. Doing that gives us:

$$y = 100 * 2^{10/5}$$

Following the order of operations, which says I have to compute the 10/5 term in the exponent next, gives me:

$$y = 100 * 2^2$$

Exponents before multiplication, so we get:

©Brian Gillispie, 2016.

$$y = 100 * 4 = 400$$

So when x is 10, y is 400.

Example 1.2.5: Evaluate $y = 2^x$ for $x = 50$

Solution to 1.2.5: This means to plug in 50 wherever you see an x. Doing that gives us:

$$y = 2^{50}$$

This simplifies to:

$$y = 1.12589997e15$$

Wait, what? You saw that one right. Most calculators will display something like the above when you try to compute it. The answer to y in Example 1.2.5 is in what is known as scientific notation, which means our answer for y really is:

$$y = 1.12589997 * 10^{15}$$

Where the power on the 10 is the number after the E on the calculator. Be sure you know how to read scientific notation when it does arise on answers you calculate, as it will occur from time to time, especially in models that run for a really long period of time.

©Brian Gillispie, 2016.

1.2.2: Issues in Evaluating Equations

There are some issues that can arise when one attempts to evaluate an equation. Here, we will mention a few of them. First, recall that it is illegal to ever divide by zero. So any equation that has you divide by zero will be undefined. However, remember that you can divide into zero.

Secondly, sometimes the value that comes out of an equation will be too big or too small for the calculator to handle properly. When that happens, we say the model has experienced overflow or underflow. The definition of those two terms is below:

> **Definition 1.2.1**: An equation has experienced **overflow** when the value of a calculation is bigger than the calculation device can store.

> **Definition 1.2.2**: An equation has experienced **underflow** when a calculation turns a term that isn't zero into zero for purposes of calculation.

Now, we will demonstrate a few Examples where division by zero, underflow, and overflow occur in the equations that we wish to evaluate.

Example 1.2.6: Evaluate the equation $y = \frac{2}{x}$ for $x = 0$

Solution to 1.2.6: Plugging in 0 for x gives us:

$$y = \frac{2}{0}$$

However, the mathematical calculation of two divided by zero is undefined. Therefore we say that y is undefined when x is 0.

©Brian Gillispie, 2016.

Example 1.2.7: Evaluate the equation $y = 2^x$ for $x = 400$

Solution to 1.2.7: Plugging in 400 for x gives us:

$$y = 2^{400}$$

However, when I attempt to compute that with my calculator, I get the message ERROR: OVERFLOW[2]. What this means is the answer is too big for my calculator to store. When that happens, we say that y is too big to compute[3] when x is 400.

Example 1.2.7 is an example of what is known as overflow, which occurs whenever the variable being stored is too big to handle. On most calculators, overflow occurs near the value $1 * 10^{100}$, though the exact value will depend on how the calculator was designed.

Example 1.2.8: Evaluate the equation $y = 1 + x$ for $x = 2.2 * 10^{-16}$

Solution to 1.2.8: Plugging in our given value for x, we end up with:

$$y = 1 + 2.2 * 10^{-16}$$

This computes to:

$$y = 1$$

Example 1.2.8 is a classic example of what is known as underflow, which occurred because the value we are adding is too small to register still when it is added to 1. Technically the answer is 1.00000000000000022, but the calculator couldn't store that properly, so it is treated as 1. Underflow tends to occur with values smaller than $1 * 10^{-16}$, though the exact point will depend on how variables are stored on your calculator.

---

[2] Some calculators will successfully compute this example, and will return $2.58 * 10^{120}$
[3] Notice we do not say undefined. The answer is defined, we just couldn't find it.

©Brian Gillispie, 2016.

Both overflow and underflow are issues you will need to be aware of when modeling. Thankfully they do not arise often, and the usually[4] do not impact the model significantly[5]. For the models we will use in this book, all we will need to know is they can occur, and even when they do, the difference between 1 and 1.00000000000000022 is insignificant for what we will be modeling.

There is one last issue a modeler needs to be aware of when evaluating equations, and that is when evaluating equations, make sure that the units of what you are plugging in matches the units of the equation. If not, you will end up with disastrous results. Let's demonstrate.

Example 1.2.9: Given the equation $y = 2200x + 4000$, and x is in days since the store opened, evaluate the equation for $x = 1$ week.

Solution to 1.2.9: Notice here that the equation wants x to be in days, and we were given x in weeks. To fix this, we need to convert our given value for x into days, which would mean[6] x is 7 when we plug it into the equation. Therefore, plugging in 7 for x gives us:

$$y = 2200(7) + 4000 = 19400$$

Therefore, when x is one week (or seven days), then y is 19400.

Always check the units you are given and be sure they match the units the equation wants. If they do not, convert them. That is always the first rule of evaluating equations. And, if no units are given, then you can safely assume that the units do not matter, and proceed like we have before.

---

[4] Be advised it really depends on what you are modeling. If you are modeling nanotechnology, underflow can be devastating.
[5] To see an example where underflow does impact a model significantly, look up catastrophic cancellation. See [13] for more on this.
[6] Remember, there are 7 days in a week.

©Brian Gillispie, 2016.

1.2 Exercises

Directions: Evaluate the given equations. If the equation results in overflow, say the answer is too big to calculate. If the equation is undefined, say undefined.

1. $y = -10x + 50$ for $x = 1$

2. $y = 0.75x + 50$ for $x = 150$

3. $y = 2x + 68$ for $x = -10$

4. $y = 10 * 2^x$ for $x = 8$

5. $y = 20 + 30 * 2^x$ for $x = 17$

6. $y = \frac{3}{x}$ for $x = 25$

7. $y = \frac{100}{x}$ for $x = 50$

8. $y = \frac{300}{x}$ for $x = 0$

9. $y = -16x^2 + 1500$ for $x = 4$

10. $y = -16x^2 + 77.65x + 10$ for $x = 2$

11. $y = -40x^2 + 102.5x + 3$ for $x = 0.6$

12. $y = -2.3x^2 + 16x + 3$ for $x = 4.2$

13. $y = x^3 + 2x^2 - 10x + 3$ for $x = 4$

©Brian Gillispie, 2016.

14. $y = -16x^2 + 77.65x + 10$ for $x = 2000$

15. $y = 2x + 45$ for $x = 2.3 * 10^{-22}$

16. $y = 3x$ for $x = 3.5 * 10^{24}$

17. $y = 1 + x + x^2$ for $x = 4.5 * 10^{-55}$

18. $y = x^x$ for $x = 42$

19. $y = 50 * 1.05^{x/7}$ for $x = 21$

20. $y = 100 * 0.97^{x/40}$ for $x = 80$

21. $P = 2L + 2W$ for $L = 100, W = 32$

22. $A = LW$ for $L = 7.6$ and $W = 5.2$

23. $A = LW$ with L and W in feet for $L = 2.7$ feet and $W = 12$ inches

24. $y = 50x + 100$ with x in dollars, for $x = 200$ cents

25. $y = -25x + 150$, with x in minutes, for $x = 0.1$ hours

26. $y = 10 * 2^{x/5}$, with x in hours, for $x = 300$ minutes

27*. $y = 10 * 2^{x/5}$, with x in hours, for $x = 525600$ minutes

©Brian Gillispie, 2016.

1.3: Dimensional Analysis

So far in this Chapter, we have discussed how models are created, as well as how to evaluate equations. Before we go any further in our discussion of modeling though, we need to discuss how to handle units and mathematical operations, which is what we will discuss next.

1.3.1: Adding and Subtracting Units

While modeling, we will often have to add or subtract two quantities that are of different units. When we do that, we need to make sure we follow the rules for how to add and subtract things of unlike units; else our answer will be a nonsense answer. Below is the rule we must follow when adding and subtracting quantities:

> **Rule for adding and subtracting:** You can only add and subtract numbers with the **same units**, and the result will be in the same units as both of the numbers.

What this means is adding and subtracting two numbers that are not in the same units is not defined. In other words, you cannot legally add 6 feet + 12 inches as both are in different units. Therefore, if you wish to add two quantities that are not in the same units, you are required to first convert them to the same units. Failure to do that will result in a nonsense answer. Let's demonstrate[7]:

Example 1.3.1: 4 hours + 3 hours = ?

Solution to 1.3.1: Both terms are in the same units, so we can legally add them. Therefore, 4 hours + 3 hours = 7 hours, and 7 hours is our final answer.

---

[7] Be advised we assume familiarity with common English conversions in this section. If you are unsure of how to convert from feet to inches or miles, or how to go from seconds to minutes to hours to days, you may wish to consult an outside source on how to do that before proceeding.

©Brian Gillispie, 2016.

Example 1.3.2: 36 inches + 6 feet = ?

Solution to 1.3.2: Both terms are not in the same units this time, so we must convert them to the same units before we can add them. Converting 36 inches[8] to feet gives us:

$$3 \text{ feet} + 6 \text{ feet} = 9 \text{ feet}$$

Therefore, our answer is 9 feet.

Notice how in the last problem the answer was not 42, but instead was 9 feet. This is why it is important to convert all units before adding, as there is a huge difference between 9 feet and 42 feet.

Example 1.3.3: 42 hours − 600 minutes = ?

Solution to 1.3.3: These terms are not in the same units, so we must convert them to the same units before proceeding. Converting 600 minutes to hours gives us:

$$42 \text{ hours} - 10 \text{ hours} = 32 \text{ hours}$$

Therefore, our answer is 32 hours.

---

[8] Note it does not matter which one you convert, as long as you end up with all of them in the same units.

©Brian Gillispie, 2016.

1.3.2: Multiplying Units

When we multiply numbers that have units, the units of the product are as follows:

> **Rule for Multiplying:** When you multiply two numbers, you have to also multiply the units.

What this means is unlike adding and subtracting, you can multiply two numbers of unlike units, but you have to also multiply the units together. Let's demonstrate how that works now with an Example.

Example 1.3.4: 17 feet * 20 feet = ?

Solution to 1.3.4: To solve this problem, you first multiply 17 and 20, which is 340. Then you also have to multiply the units of 17 and 20, which means you have to compute feet * feet, which is feet^2. Putting this together gives us our final answer of:

$$340 \text{ feet}^2$$

Notice how we had to take feet * feet as well in the problem. Since feet is only a unit, all we could do with that is simplify it down to feet^2, by the rules of exponents[9].

Example 1.3.5: 12 feet * 36 inches = ?

Solution to 1.3.5: Notice that this time the two numbers are not in the same units. However, we can proceed with them like this by computing 12 times 36 and feet times inches. If we do that, we will end up with:

---

[9] There is a brief review of rules of exponents in Chapter 5, but basically we had two of the same quantity being multiplied, which is, by definition, that same quantity squared.

©Brian Gillispie, 2016.

$$432 \text{ (feet} * \text{inches)}$$

Which is read as 432 feet-inches[10]. However, we don't tend to measure things in feet-inches, and if you went into the store and tried to buy 432 feet-inches of carpet, they would probably be utterly confused as to what you are talking about. So, for that reason, it is more common to convert to the same units before we multiply (when possible), so that we end up with units that make more practical sense. If we do that in Example 1.3.5, we end up with:

$$12 \text{ feet} * 3 \text{ feet} = 36 \text{ feet}^2$$

Which would result in a final answer of 36 feet^2.

The interesting thing about Example 1.3.5 is both answers listed are correct. However, one of them is useful and the other is not. When working problems involving multiplication of units, you prefer to end up with units in your final answer that are practical, and match how everyone else measures things. We measure things in square feet[11]. We do not measure things in feet-inches. For that reason, if you see that multiplying the numbers as is will result in units that are not practical to use, consider converting everything to the same units first, then multiply[12].

---

[10] That may look like feet minus inches, but that is actually a hyphen. It is common to hyphenate when you have two unlike units that are multiplied. For another example of this, look at an electric bill, as usage is in kilowatt-hours, which came from kilowatts * hours.
[11] For example, carpet.
[12] If unsure, remember the standard is square units mean area and surface area and cubic units mean volume.

©Brian Gillispie, 2016.

1.3.3: Dividing Units

When we divide two numbers that involve units, the units of the division are as follows:

Rule for Dividing: When you divide two numbers, you have to also divide the units.

This means that the units of the final answer will be the result when you divide the respective units of the numbers being divided. Let's demonstrate.

Example 1.3.6: $\frac{12 \text{ feet}}{2 \text{ feet}} = ?$

Solution to 1.3.6: To solve this problem, we first divide 12 by 2, and get 6. Then we divide feet by feet, which is:

$$6 * \frac{\text{feet}}{\text{feet}}$$

However, feet over feet cancel each other out, which means the answer is just 6. No units exist in the final answer.

Example 1.3.7: $\frac{70 \text{ cookies}}{14 \text{ guest}} = ?$

Solution to 1.3.7: First, we divide 70 by 14, which is 5. Then we have to divide cookies by guest. However, we cannot work out that division, so we leave it in our answer as:

$$5 \frac{\text{cookies}}{\text{guest}}$$

©Brian Gillispie, 2016.

Notice how the units are written in fraction form first, with cookies on top and guest on the bottom, and that is because we were required to divide cookies by guest, but since we couldn't, we wrote it as a fraction. Therefore, our final answer is:

$$5 \frac{\text{cookies}}{\text{guest}}$$

Which is often read as 5 cookies per guest.

Example 1.3.8: $\frac{280 \text{ miles}}{70 \frac{\text{miles}}{\text{hours}}} = ?$

Solution to 1.3.8: First, we divide 280 by 70, which is 4. Then we have to divide miles by $\frac{\text{miles}}{\text{hours}}$, which is:

$$\text{miles} \div \frac{\text{miles}}{\text{hours}}$$

Remember that to divide by a fraction you invert the fraction you are dividing by and change it to a multiplication problem, which results in:

$$\text{miles} * \frac{\text{hours}}{\text{miles}}$$

The two miles cancel each other out, resulting in hours for our final units. Therefore, the final answer is:

$$4 \text{ hours}$$

In Example 1.3.8, we were using the classic distance-rate-time equation, for the case where someone drives 280 miles at 70 miles per hour, and solved for how long it takes them to drive to that destination. Notice that our final answer also worked out to be the proper units for

©Brian Gillispie, 2016.

the answer, which is hours. If you remember these rules for how to add, subtract, multiply, and divide problems with units, you can then use them to check and see if your answer makes sense and is in the proper units. For example:

Example 1.3.9: While working with the simple interest equation $I = Prt$, you plugged in $10000, r = \frac{0.05}{\text{year}}$ and $t = 90$ days and came up with the answer that this investment earns $I = 10000 * 0.05 * 90 = \$450$. Are the units correct for this answer?

Solution to 1.3.9: Let's check the multiplication and see if the units are what we claim they are, which is dollars. We have the following units being multiplied:

$$I = \$10000 * \frac{0.05}{\text{year}} * 90 \text{ days}$$

Now, it is true that $10000 * 0.05 * 90$ is 450. However, when we check our units, since the units have to be multiplied, that means we needed to multiply the following for our units of our final answer:

$$\$ * \frac{1}{\text{year}} * \text{days}$$

Which, via the rules for multiplying fractions, is:

$$\frac{\$ * \text{days}}{\text{hours}}$$

So, our answer for those numbers should have been:

$$I = \frac{\$450 * \text{days}}{\text{hours}}$$

©Brian Gillispie, 2016.

That happens to be a nonsense unit, which probably means something was not plugged in correctly. And, in fact, that was the case here, as the simple interest formula requires that t be in years, not days. By checking the units, we were able to verify that an error had been made in the process of plugging in the numbers, which would allow us to then go back and fix the error, and then, hopefully acquire the correct answer.

1.3 Exercises

Convert all numbers to like units when possible, and then work the following calculations.

1. 60 minutes + 90 minutes = ?

2. 3 miles + 12 miles = ?

3. 4 hours − 2 hours = ?

4. 2 inches + 7 inches − 3 inches = ?

5. 360 seconds + 4 minutes = ?

6. 24 hours + 3 days = ?

7. 10 days − 36 hours = ?

8. 10 yards − 15 feet = ?

9. 24 inches + 3 feet + 2 yards = ?

10. 14 days + 12 hours − 600 minutes = ?

11. 3 feet * 10 feet = ?

12. 110 inches * 46 inches = ?

13. 2 miles * 4 miles = ?

©Brian Gillispie, 2016.

There is no conversion possible between these units. Calculate them appropriately.

14. $\dfrac{200 \text{ yards}}{10 \text{ cows}} = ?$

15. $\dfrac{130 \text{ cars}}{4 \text{ stoplights}} = ?$

16. $\dfrac{48 \text{ pizza slices}}{12 \text{ people}} = ?$

17. $\dfrac{350 \text{ miles}}{50 \frac{\text{miles}}{\text{hours}}} = ?$

18. $\$600 * \dfrac{2 \text{ cookies}}{\$} = ?$

19. $\$150 * \dfrac{1 \text{ pizza}}{\$10} = ?$

20. $\$6 * \dfrac{4 \text{ ramen packages}}{\$1} = ?$

21. $\dfrac{\$8.99}{\text{order}} * 5 \text{ orders} = ?$

22. $\dfrac{\$25}{\text{item made}} * 400 \text{ items made} = ?$

23. $\dfrac{\$600}{\text{computer}} * 14 \text{ computers} = ?$

©Brian Gillispie, 2016.

1.4: Errors

In the process of using a model, errors can and will occur. In this Section, we will discuss and define error, as well as discuss the different types of ways that error can occur when creating and using a model.

1.4.1: Measures of Error

To begin, we will start out by defining what we mean by error. An error in a calculation is defined as follows:

> *Definition 1.4.1:* The error in a mathematical calculation is defined as the distance between the real value and the calculated value, which is:
>
> $$Error = calculated\ value - actual\ value$$

Let's now demonstrate this definition with a few Examples before proceeding.

Example 1.4.1: The scale at the vet claims your cat weighs 15 pounds, but in reality the cat weighs 14 pounds. What is the error in this measurement?

Solution to 1.4.1: Our actual value is 14 pounds, and the measured value is 15 pounds. This gives us an error of the following:

$$Error = 15 - 14 = 1\ pound$$

Example 1.4.2: A tumor in the brain is measured at 0.20 grams, but when it is removed it is really 0.30 grams. What was the error in the measurement?

©Brian Gillispie, 2016.

Solution to 1.4.2: In this case, our actual was the real value of 0.30 grams, and our calculated value was the measured value of 0.20 grams, which gives the following for the error:

$$\text{Error} = 0.20 - 0.30 = -0.10 \text{ grams}$$

As you can see in the Examples, error can be positive or negative. However, usually people refer to the error of a measurement without the sign, so in Example 1.3.2, most would say there was an error in the measurement of 0.1 grams, not -0.1 grams[13].

While reading these Examples you may have realized that there is more to consider than just the amount of the error. For instance, most people don't care that their cat was weighed incorrectly, but being too far off in the weight of a brain tumor can be fatal to the person the tumor is in! For this reason, modelers also calculate something called the relative error, which is used to determine by what percent the measurement was off by. The relative error of a measurement is defined as follows:

Definition 1.4.2: Relative error is defined as the percent the error is of the actual amount. It is computed by the following formula:

$$\text{Relative Error} = \frac{\text{calculated value} - \text{actual value}}{\text{actual value}} * 100$$

Just like with error, most people are interested in the amount of the relative error, and not the sign of the relative error. Let's now demonstrate how to calculate the relative error for our previous two Examples.

---

[13] When you refer to the error without the sign, you are referring to what is called the absolute error.

©Brian Gillispie, 2016.

Example 1.4.3: A tumor in the brain was measured to be 0.20 grams, but when it is removed it is really 0.30 grams. What was the relative error in the measurement?

Solution to 1.4.3: Remember from Example 1.4.2 that the actual value was 0.30 grams, and the calculated value was 0.20 grams. Plugging those into the equation for relative error gives us:

$$Relative\ Error = \frac{0.20 - 0.30}{0.30} * 100$$

Or, a relative error of -33.33% once calculated.

Example 1.4.4: The scale at the vet claims your cat weighs 15 pounds, but in reality the cat weighs 14 pounds. What is the relative error in this measurement?

Solution to 1.4.4: Recall from Example 1.4.1 that the actual value was 14 pounds, and the calculated value was 15 pounds. Plugging those into the equation for relative error gives us:

$$Relative\ Error = \frac{15 - 14}{14} * 100$$

Or, a relative error of -7.14%, rounded to the nearest hundredth of a percent.

As you can see, the measurement of the tumor had a higher relative error, but a lower error, but the error in the measurement of the cat had a higher error, but a lower relative error. When we model, we usually desire to keep the relative error as low as possible, not the actual error. However, this is not without flaws, as the next Example will demonstrate.

Example 1.4.5: The weather reporter reported the temperature outside as being 1 degree, but it is really 0.01 degrees. What is the relative error in this temperature report?

Solution to 1.4.5: In this example, the calculated error was 1 degree, and the actual value is 0.01 degrees. Therefore, the relative error is:

©Brian Gillispie, 2016.

$$\frac{1-0.01}{0.01} * 100 = \frac{0.99}{0.01} * 100 = 9900$$

Therefore, the relative error in this forecast is 9900%.

Now, in this last example, the temperature forecast was off not even one full degree, yet the relative error was a whopping 9900%! This would say that the forecast was way off, but in reality, an error of 0.99 degrees is not a serious error. In fact, in both cases, you would still wear your coat outside. So it is not like the error was the difference between you wearing a swimsuit and you wearing a coat. This however demonstrates the flaw in depending too much on the relative error, and that is if the actual value is close to 0, the relative error will skyrocket! Therefore, for that reason, relative error is best used on values that are nowhere near zero, just to avoid that issue.

Another issue with relative error is that it will not catch underflow error. If for example the calculated value was 1 due to underflow, but the actual value was $1 + 2.2 * 10^{-16}$, the relative error equation will return 0, even though there was error present in the calculation. There is nothing that can be done to fix this, except to not use the relative error equation to calculate the error of numbers that are really close together.

©Brian Gillispie, 2016.

1.4.2: Types of Errors

Now that we have discussed how to calculate error, we will now take a look at the some of the various types of errors that can occur in modeling. There are three common types of errors that one has to watch out for when modeling. These are systematic errors, random errors, and rounding errors. Let's start by defining what is meant by a systematic error in a model:

> *Definition 1.4.3:* An error is systematic if the error is applied uniformly to all the measurements made

What this means is for an error to be systematic, the error has to be applied equally to everything measured. One example of this would be a scale that says everything that it weighs is 1 pound heavier than it really is. Another example of this would be a thermometer that says the temperature is really 5 degrees cooler than it really is.

Usually systematic errors are caused by faulty measuring devices that need to be recalibrated. Checking to make sure that all measuring devices are measuring correctly can reduce the likelihood of a systematic error occurring.

The second type of error is random error. A random error is defined as follows:

> *Definition 1.4.4:* An error is random if the error is caused by uncontrollable and unexpected factors.

While the definition is pretty broad, finding random errors is not hard. A random error could be something as simple as a misreading of a rain gauge, like for instance the rain gauge is really holding 8.75 inches but you read it as 9 inches of rain. Or, the random error could occur because the object you are trying to weigh will not sit still, like a young baby.

Unfortunately there is nothing that can be done to prevent random errors, except to be as careful as possible. Thankfully most random errors are by small amounts and will not usually mess up a model too significantly.

©Brian Gillispie, 2016.

The third type of error is rounding error. At first this error may not seem as obvious. After all, we round numbers all of the time! Even when we computed our relative errors we rounded to the nearest hundredth. And yet, despite that, rounding does create error. How major that error is depends on what you do with the rounded number. Let's demonstrate what we mean with a couple Examples:

Example 1.4.6: Convert 68 degrees Fahrenheit to Celsius. Recall that the formula to convert from Fahrenheit to Celsius is given by $C = \frac{5}{9}(F - 32)$

Solution to 1.4.6: For this problem, we are going to round $\frac{5}{9}$ to $.56$, which is the fraction rounded to the nearest hundredth. With that change, the problem becomes the following:

$$C = .56 * (68 - 32)$$
$$C = .56 * 36$$
$$C = 20.16 \text{ degrees}$$

However, the real answer, if you use $\frac{5}{9}$ as a fraction and do the proper fraction math, is 20 degrees. Using the rounded answer of $.56$ introduced a rounding error of 0.16 degrees.

Now, an error of 0.16 degrees does not appear significant. However, whether or not it is significant depends on what you wish to do with the result! If you are doing this calculation to see whether to wear a coat outside, 0.16 degrees makes no difference. However, if you are doing this conversion to calculate a temperature to then plug into the model, being off by 0.16 degrees can make a huge difference. To see why, let's look at the next Example:

Example 1.4.7: You are working on a model to predict something, and part of your model says to take the previously computed value, square it, and then subtract 2. This new number is plugged back into the equation and repeated 14 times[14]. The real starting value is 0.51 when the

---

[14] This is an example of a discrete sequence, which we will discuss in Chapter 4.

©Brian Gillispie, 2016.

model starts, but you decide to round to 0.50, figuring how much can 0.01 matter in a calculation. The results of the calculation for each step are as follows:

Approximate Value	Real Value
0.500000	0.510000
-1.750000	-1.739900
1.062500	1.027252
-0.871094	-0.944753
-1.241196	-1.107441
-0.459433	-0.773574
-1.788921	-1.401583
1.200239	-0.035564
-0.559427	-1.998735
-1.687041	1.994942
0.846107	1.979795
-1.284103	1.919587
-0.351080	1.684815
-1.876743	0.838601

Look at the last line of this calculation. Our calculation with the approximate value came up with -1.876743, and the real value has 0.838601. This is an error of -2.715344, and a relative error of -323.79%! So the 'insignificant' rounding off of 0.01 caused a huge error after only 14 runs of a model.

The moral of the story is this: A small change in a starting value to a problem can have widespread implications on future calculations. Those 'minor' rounding steps conducted throughout the problem can and often do cause widely different final answers to occur. For Example, rounding the number 0.0047 to 0.005 in a calculation of the value of a house can[15] cause an error in the final value of the house of around $8,000! Therefore, to avoid errors like this from occurring, it is best to avoid rounding until the very end of a problem.

---

[15] To see this, divide 650 by both numbers, and see by how much they are off. Those two numbers are the difference in the value of the house.

©Brian Gillispie, 2016.

1.4.3: Error Tables

When a mathematical model is used, it needs to be analyzed against real world data to see how well the model represents the real world data. This is often done by calculating the error of the model with all of the available data points. This is often done by using what is known as an error table, where we list the error of the model for each data point, as well as the total absolute error of the model.

In order to make an error table, start by listing all the of the known data points in the table. Once the data points are listed, calculate the error for that data point, and note the error on the same row as the point is listed on. Once that is done, drop the signs of the error, and add them up. Let's demonstrate this with an Example.

Example 1.4.7: Given the data points $(0, 0)$, $(1, 61)$, $(2, 78)$, $(3, 61)$, and $(4, 0)$, calculate an error table for the model $y = -16x^2 + 72x$

Solution to 1.4.7: Start out by listing the data points[16], each data point in its own row, as follows:

X	Y
0	0
1	61
2	78
3	61
4	0

Next, compute the error for each data point. Calculate the error by plugging in the x value into the model, to acquire the calculated value. Once you have the calculated value, then compare the calculated value with the actual value, which is the y value on that row of the table. The calculated value for the x values for each of the data points are now listed by the appropriate x values in the table, and the table will now look like the following:

---

[16] If unstated, data points are assumed to be in the format (x, y)

©Brian Gillispie, 2016.

X	Y	Calculated Value
0	0	0
1	61	56
2	78	80
3	61	56
4	0	0

Next, calculate the error, by subtracting the calculated value from the actual value, which is y. Doing this gives us the following for our table:

X	Y	Calculated Value	Error
0	0	0	0
1	61	56	−5
2	78	80	2
3	61	56	−5
4	0	0	0

Notice that the error is the calculated value minus the y value in each row. So in row one, we have $0 - 0 = 0$ for the error. On row two, we have $56 - 61$, which gives us -5 for the error. Continuing like this gives us the error as listed in our table. Once the error is listed, drop the signs and add up the numbers in the error column to get the sum of the errors, as such:

X	Y	Calculated Value	Error
0	0	0	0
1	61	56	5
2	78	80	2
3	61	56	5
4	0	0	0
Sum			12

This tells us that when x is 0 and 4, the model has no error, and the sum of the absolute error of the model is 12.

©Brian Gillispie, 2016.

Example 1.4.8: Given the model $y = 100 * 1.05^{x/5}$ and the data points $(0, 100)$, $(5, 106)$, $(10, 113)$, $(15, 125)$, $(20, 145)$, and $(25, 170)$, calculate an error table for these points and the given model.

Solution to 1.4.8: Start by creating the error table, remembering that our given data points go in the x and y columns:

X	Y	Calculated Value	Error
0	100		
5	106		
10	113		
15	125		
20	145		
25	170		

Next, acquire the calculated value by plugging in the x value in each row into the model. Once you do this, the table will now become:

X	Y	Calculated Value	Error
0	100	100	
5	106	105	
10	113	110.25	
15	125	115.7625	
20	145	121.550625	
25	170	127.6281563	

Finally, compute the error by calculating the calculated value minus the y value for each row. Once that is done, the table becomes:

X	Y	Calculated Value	Error
0	100	100	0
5	106	105	−1
10	113	110.25	−3.25
15	125	115.7625	−9.2375
20	145	121.550625	−23.449375
25	170	127.6281563	−42.3718437

©Brian Gillispie, 2016.

Finally, drop the signs in the error column, and add up the numbers in that column. Once that is done, we will have the following:

X	Y	Calculated Value	Error
0	100	100	0
5	106	105	1
10	113	110.25	3.25
15	125	115.7625	9.2375
20	145	121.550625	23.449375
25	170	127.6281563	42.3718437
Sum			79.3087187

Therefore, the total sum of the absolute error of this model is 79.3087187

When mathematical models are created, the goal is (usually) to create a model that has the lowest possible error for the given data points. If possible, we would like to create a model that has no error whatsoever with the given data points, which would mean that every entry in the error column in the table would be zero. However, this is often not possible, so instead we often will create multiple models, then select the one with the lowest error to use. Let's close out this Section with an example where we will compare two models and their sum of errors[17]:

Example 1.4.9: Given the data points $(0, 100), (5, 106), (10, 113), (15, 125), (20, 145)$, and $(25, 170)$, compare the two models $y = 100 * 1.05^x$ and $y = 2.5x + 100$

Solution to 1.4.9: We already calculated the error table for the first model in Example 1.4.8, but will include it here for easy reference. The error table, with the sign of the error discarded is:

---

[17] Note that the sum of errors is not the only selection criterion used. Many use the sum of the squares of the errors, for reasons we will discuss much later in the book.

X	Y	Calculated Value	Error
0	100	100	0
5	106	105	1
10	113	110.25	3.25
15	125	115.7625	9.2375
20	145	121.550625	23.449375
25	170	127.6281563	42.3718437
Sum			79.3087187

Likewise, the final error table for the model $y = 2.5x + 100$ is as follows[18], with the signs dropped for the error:

X	Y	Calculated Value	Error
0	100	100	0
5	106	112.5	6.5
10	113	125	12
15	125	137.5	12.5
20	145	150	5
25	170	162.5	7.5
Sum			43.5

Therefore, since the sum of the errors for the model $y = 2.5x + 100$ is lower, this would be the best model to use for modeling these data points.

---

[18] We omitted the steps of calculating the error table this time! See if you can get the same answer we did.

©Brian Gillispie, 2016.

1.4 Exercises

For Exercises 1 - 6, calculate the error and the relative error

1. You buy a bag of potato chips that says it has 8 ounces of chips in it, but you actually have 7.9 ounces of potato chips

2. A scale at the doctor's office says you weigh 175 pounds, but your real weight is 172 pounds.

3. Your internet provider claims to provide download speeds of 15 MB, but you really are getting 16 MB download speeds.

4. Your new car claims to get 43 mpg, but it really gets 40 mpg.

5. The oven says it is 350 degrees inside, but it is really 348 degrees.

6. Radar says you are driving 65 mph, but you were really going 68 mph.

For Exercises 7 - 12, state whether the error is systematic, random, or rounding, and why you pick that answer

7. An altimeter for a plane says the plane is really 500 feet more off the ground than it really is

8. While weighing yourself on the scale, the cat steps on it without your knowledge.

9. A thermometer is placed in the shade at the airport in Las Vegas

10. While calculating how many mpg your car gets, you decide to report 236.2 miles as 236 miles

11. A football quarterback is always 1 foot off target on all of his passes during a game.

©Brian Gillispie, 2016.

12. A school nurse is taking temperatures with a thermometer that is reporting temperatures that are 1 degree higher than they really are.

13* Look up chaos theory on the internet and write a small report on it, answering the following questions in your report: Where did chaos theory originate from? What are the implications of this theory on mathematical modeling?

14. Create an error table for the data points $(0, 5), (1, 9), (2, 16), (3, 25)$, with the model $y = 5x + 5$

15. Create an error table for the data points $(0, 0), (1, 34), (2, 28), (3, 0)$, with the model $y = -16x^2 + 48x$

16. Create an error table for the data points $(0, 10), (1, 25), (2, 35), (4, 45)$, with the model $y = 10x + 10$

17. Create an error table for the data points $(1, 10), (2, 4.5), (4, 2)$ with the model $y = \frac{10}{x}$

18. Create an error table for the data points $(0, 50), (1, 36), (2, 25), (3, 20)$ with the model $y = -10x + 50$

19. Create an error table for the data points $(0, 100), (5, 130), (10, 168), (15, 206), (20, 233)$ with the model $y = 100 * 1.3^{x/5}$

20. Given the data points $(0, 15), (1, 30), (2, 55), (3, 110)$ which of these two models has the lowest error? $y = 15x + 15$ and $y = 15 * 2^x$

©Brian Gillispie, 2016.

21. Given the data points $(0, 100), (1, 132), (2, 166), (3, 132)$ which of these two models has the lowest error? $y = -16x^2 + 48x + 100$ and $y = 100 * 1.25^x$

22. Given the data points $(1, 20), (2, 9), (3, 4), (4, 2)$ which of these two models has the lowest error? $y = \frac{20}{x}$ and $y = 40 * 0.5^x$

23. Given the data points $(0, 100), (1, 110), (2, 119), (3, 126), (4, 133)$, which of these three models has the lowest error? $y = 10x + 100$, $y = 100 * 1.08^x$ and $y = 9x + 103$

24. Given the data points $(0, 150), (1, 188), (2, 220), (3, 250)$, which of these three models has the lowest error? $y = 150 * 1.25^x$, $y = 160 * 1.2^x$ and $y = 165 * 1.15^x$

25. Given the data points $(0, 50), (1, 46), (2, 40), (3, 32)$ which of these three models has the lowest error? $y = -6x + 50$, $y = -5x + 48$ and $y = -5.5x + 49$

©Brian Gillispie, 2016.

1.5: Application: Some Common Models

Now that we have discussed a few of the basics involved in creating and using models, we are going to discuss and use a few well known models that are already in use in the real world. These models are the models used to measure area and perimeter, as well as the distance-rate-time model.

1.5.1: Measuring Space: Area and Perimeter

Oftentimes, people wish to figure out how much space an object will take up. Two of the ways of measuring space are the perimeter and the area[19]. Perimeter is the measurement of how much space it takes to enclose an object, and area is the measurement of how much space the object itself takes up. Another way to think of it is, if you had a lot of chocolate, perimeter would be the measurement of how big the fence is you would put around it, and area is how much ground space it takes to place all your chocolate on the ground.

Most of the common shapes in the world have a well-known perimeter and area formula. A few of the more common ones are below[20]:

Rectangle:
   Area: LW, *Perimeter:* $2L + 2W$

Circle:
   Area: πr^2, *Perimeter:* $2\pi r$

Triangle:
   Area: $0.5bh$, *Perimeter:* $s_1 + s_2 + s_3$

---

[19] You can also measure the volume of an object, which is how much three dimensional space the object takes. We will, however, focus on area and perimeter in this section.
[20] Be advised that it is mathematically impossible to include a perimeter and area formula for every shape in existence. Therefore, it is possible you may run into a shape not mentioned here.

©Brian Gillispie, 2016.

For these formulas, L stands for Length, W stands for Width, r stands for radius, b stands for base, h stands for height, and s_1, s_2 and s_3 are the first, second and third sides of the triangle. Also note that for any of these formulas to work, everything you plug in must be in the same units, else you will either end up committing an illegal operation when you add, or will end up with nonsense units in your final answer. Always convert everything to the same units before working an area or perimeter problem.

Now, let's demonstrate how to use these formulas with a few Examples.

Example 1.5.1: Find the area of a rectangle with a width of 20 feet and a length of 25 feet.

Solution to 1.5.1: Since length and width are in the same units, we can directly plug into the formula. Per the table at the start of this Section, the formula for the area of a rectangle is LW, so plugging in our given values we end up with:

$$25(20) = 500$$

Also, since we multiplied two units which were both in feet, our final answer is in feet * feet, or feet^2. Therefore, the area of this rectangle is 500 feet^2.

Example 1.5.2: Find the perimeter[21] of a circle with a radius of 6 inches.

Solution to 1.5.2: Per the table at the start of this Section, the perimeter of a circle is $2\pi r$. Plugging in our given value for the radius of the circle into r gives us:

$$2\pi(6) = 12\pi$$

Also, since our units for the radius were in inches, we need to multiply inches by whatever units 2 is in. However, 2 was just a number, and it had no units. Therefore, our final

---

[21] The perimeter of a circle is often called the circumference of the circle. We'll use perimeter though for consistency with all the other shapes.

©Brian Gillispie, 2016.

answer is still in inches, as inches times something not in units is still in inches. This means our final answer is then[22]:

$$12\pi \text{ inches}$$

Example 1.5.3: Find the perimeter of a triangle with sides of 10 inches, 1 foot, and 2 yards.

Solution to 1.5.3: This time we cannot directly plug in our given numbers, as they are all in different units. Plus, even if we did plug them in, it is illegal to add 10 inches + 1 foot + 2 yards. Therefore, we need to convert all of these measurements to the same units. Which unit we convert them to doesn't matter, just make sure all three of them are in the same units. We'll convert all of these to inches, and since there are 12 inches in a foot and 36 inches in a yard, that means our sides are now 10 inches, 12 inches, and 72 inches respectively.

Now that all the measurements are in the same units, we can plug them into the perimeter equation and solve. The perimeter equation for this shape (a triangle) is $s_1 + s_2 + s_3$. Plugging in those three sides into the equation gives us that our perimeter is:

$$10 + 12 + 72 = 94$$

Since all of our units are in inches, our final answer is in inches[23], which means our final perimeter is 94 inches.

Perimeter and area measurements can be useful when trying to figure out how much fencing to buy, or how much carpet to buy to carpet a room. A proper measurement will allow you to make sure you have enough money on hand to buy the amount you need, and not end up short when at the register. Let's end this Section on perimeter and area by working an Example where we try to estimate costs.

---

[22] Note that you could evaluate π, since $\pi \approx 3.14.159$. However, since we wish to avoid introducing roundoff error in our answers, we will leave the symbol π in our answers and not evaluate it.

[23] See Section 1.3 for more on the rules when we add things with units if unsure why this is.

©Brian Gillispie, 2016.

Example 1.5.4: You wish to carpet a room which is 12 feet by 8 feet. Each square foot of carpet costs $1.14. How much will it cost you to carpet this room?

Solution to 1.5.4: In order to answer this, we need to know how many square feet of carpet we need, which means we need to compute the area. Since area is LW, and, since both of our measurements are in feet, we can safely plug them into our area equation and not end up with a nonsense unit in the end. Doing that results in:

$$LW = 12 \text{ feet} * 8 \text{ feet} = 96 \text{ feet}^2$$

This means we need 96 square feet to carpet our room. Since carpet is $1.14 per square foot, this means the total cost for carpeting the room is:

$$96 \text{ feet}^2 * \frac{\$1.14}{\text{feet}^2} = \$109.44$$

Therefore, it will cost you $109.44 to carpet this room, based on those figures[24].

---

[24] Side note: In reality it would be a little higher, for two reasons. First you would have to pay the sales tax on that purchase. And secondly, no one buys exactly the amount of material they need, they always buy a little over, in case of cutting or fitting errors. This extra material would also add into the final cost. Still, this figure gives you a good starting estimate.

©Brian Gillispie, 2016.

1.5.2: Distance-Rate-Time

One of the more well-known algebra equations is the famous distance-rate-time equation, which is of the form:

$$Time\ to\ arrive\ at\ a\ destination = \frac{Distance\ to\ Travel}{Speed\ you\ are\ traveling\ at}$$

This is often shorted to:

$$T = \frac{D}{S}$$

For this equation to work, the distance and speed need to be measured in the appropriate units. So if distance is in the units a, then speed must be in the units a per b, and time will then be in the units b. For example, if distance is in miles, and speed is in miles per hour, then time will be in hours (miles took the place of a, hour the place of b). If things do not match up like that, you will have to convert units before using the equation[25].

Let's now demonstrate this formula with a few Examples.

Example 1.5.5: You wish to drive to a city which is 240 miles away. If you can drive 65 miles per hour the entire way, how long will it take you to reach your destination?

Solution to 1.5.5: In this problem, everything is in the appropriate units, so we can use the distance-rate-time equation. Plugging in our givens into the equation gives us:

$$T = \frac{240}{65} = 3.69\ hours$$

Therefore, it will take us 3.69 hours to reach our destination.

---

[25] Thankfully, the units almost always match when this formula is used. Still, check and make sure time is in the appropriate units when done, do not assume it always is.

©Brian Gillispie, 2016.

Example 1.5.6: Your daily commute to work is 10 miles, and you are able to average 70 miles per hour the entire way. How long does your daily commute take?

Solution to 1.5.6: Everything is in the appropriate units again, so we can use the distance-rate-time equation. Plugging in our givens into the equation gives us:

$$T = \frac{10}{70} = 0.14 \text{ hours}$$

Example 1.5.7: You have decided to take up running! Given that you have been running for 15 minutes, and you are running at the speed of 250 feet per minute, how far have you ran?

Solution to 1.5.7: This time we are given the time and the speed, and time and speed are in the proper units, so we can plug them into the equation. Doing that gives us:

$$15 = \frac{D}{250}$$

To solve for D, multiply both sides by 250. Doing that results in:

$$15 \text{ minutes} * 250 \frac{\text{feet}}{\text{minute}} = D$$

Using the rules from Section 1.3 for multiplying units gives us a final answer of:

$$D = 3750 \text{ feet}$$

©Brian Gillispie, 2016.

Example 1.5.8: You live 25 miles away from work, and wish to be at work in 0.5 hours. What speed do you need to drive for this to happen?

Solution to 1.5.7: In this problem, we know that our distance is 25 miles, and our time is 0.5 hours. Plugging both of those into our equation gives us:

$$0.5 = \frac{25}{r}$$

Now we need to solve the equation for r. First, multiply every side of the equation by r. That gives us:

$$0.5r = 25$$

Divide both sides by 0.5 to end up with:

$$r = 50$$

Finally, figure out what units this answer is in. Since 25 was in miles, and 0.5 was in hours, and we divided 25 by 0.5, this means our answer is in the units miles per hour. Therefore, to get to work in 0.5 hours, you need to average at least 50 miles per hour or more during the drive.

©Brian Gillispie, 2016.

1.5 Exercises

Problems with a M by them use Section 1.5.1, and problems with a D use Section 1.5.2

M1: What is the area of a rectangle with a length of 18 inches, and a width of 22 inches?

M2: What is the perimeter of a rectangle with a length of 18 inches, and a width of 22 inches?

M3: What is the area of a circle with a radius of 14 feet?

M4: What is the area of a triangle with a base of 11 feet, and a height of 100 feet?

M5: What is the area of a rectangle with a length of 10 feet, and a width of 24 inches?

M6: What is the perimeter of a triangle with lengths of 5 feet, 7 feet, and 18 inches?

M7: What is the area of a circle with a radius of 3 miles?

M8: What is the perimeter of a rectangle with a length of 150 feet, and a width of 6 inches?

M9: What is the area of a rectangle with a length of 150 feet and a width of 6 inches?

M10: What is the area of a triangle with a base of 25 feet, and a height of 15 yards?

M11: You wish to carpet a room that is 10 feet by 15 feet. The cost of the carpet is $0.90 per square foot. How much will it cost to carpet this room?

M12*: Using the same data as in Exercise 1.5.M11, find out how much it costs to carpet the room if you decide to buy 10% more carpet than needed to allow for waste, and also have to pay 5% tax on the entire purchase.

©Brian Gillispie, 2016.

M13: You wish to paint a wall that is 9 feet high by 25 feet wide. Paint is sold in containers which cost $10 each, but each container paints 100 square feet. How much will it cost you to paint this wall? *Hint: Be careful! You cannot buy partial containers of paint!*

M14: You planted corn all over your farm, which is 1.5 miles by 1.2 miles. However, now it is time to pick the corn, and you have hired help to help you pick the corn. If each helper can pick 200 square feet of corn in an hour, and we assume each helper is paid $9 per hour, how much will it cost you to have all of your corn picked? *Hint: Convert the size of the farm to feet before starting.*

M15*: Using the same data as in Exercise 1.5.M14, if we assume that each square foot of corn produces 10 ears of corn, what is the lowest you can sell each ear of corn for and still make a profit, after paying all the helpers?

D1: You wish to go on a long road trip to a destination that is 2000 miles away. If we assume that you drive 50 miles per hour the entire way, how long will it take you to reach your destination?

D2: You decide to walk to your friend's house which is 3500 feet away. If you can walk 150 feet a minute, how long will it take you to talk to your friend's house?

D3: A runner is running a 100 meter dash! If the runner can run 9.6 meters per second, how long will it take them to finish the race?

D4: A snail is moving at the speed 3 feet per minute. How long will it take this snail to reach a location that is 5280 feet away?

D5: You decide to drive all around the USA. If you are planning on driving 10000 miles at 65 miles per hour, how long will you spend on this trip just driving?

©Brian Gillispie, 2016.

D6: You decide to drive to a sports event that is 150 miles away. How long will it take you to get there at 55 miles per hour?

D7: Using the same data as in Exercise 1.5.D6, how much time will you save if you drive 65 miles per hour instead of 55 miles per hour?

D8: You wish to drive to a place that is 260 miles away, and you wish to be there in 4.25 hours or less. What is the slowest speed you can drive to make this happen, in miles per hour?

D9: You wish to run the 100 meter dash in 12 seconds or less. What is the slowest speed you can run to make this happen, in meters per second?

D10: You decide to walk to your friend's house, which is 1000 feet away, and you wish to be there in 8 minutes or less. What is the slowest speed you can walk to make this happen, in feet per minute?

D11: You decide to go on a walk. Given that you have been walking for 20 minutes, and you are walking at the speed of 110 feet per minute, how far have you walked?

D12: You decide to cruise around town randomly. Given that you have been driving at 25 miles per hour, and you have been driving for 0.65 hours, how far have you driven?

D13: You decide to take up running! Given that you have been running for 3 hours, and you have been running at the speed of 2.4 miles per hour, how far have you ran?

D14: You decide to go on a long bike ride! Give that you have been biking for 2.3 hours, and you have been biking at the speed of 17.6 miles per hour, how far did you bike?

©Brian Gillispie, 2016.

Chapter 1 Summary

1.1: In this Section, we covered the modeling cycle. To create a model, first we collect data from the real world. Once we have the real world data, we make simplifying assumptions in order to make sure that the model is able to be used practically. Once the assumptions are made, the model is created, then tested against real world data. After testing the model against real world data, the model is adjusted as necessary, and the cycle repeats.

1.2: In this Section, we defined what it means to evaluate an equation, as well as discussed the concept of overflow and underflow, and when they can arise in modeling.

1.3: In this Section, we discussed how to handle units when we add, subtract, multiply, or divide. When we add or subtract, everything must be in the same units, or the operation cannot be performed. And, when we multiply or divide, we also have to multiply or divide the units.

1.4: In this Section we defined error, as well as how it is measured. The key calculations were as follows:

$$error = calculated - actual$$

$$relative\ error = \frac{calculated - actual}{actual}$$

We also defined systematic, random, and rounding errors. Systematic are errors that are evenly applied to all calculations, random errors are errors that are unpredictable, and rounding errors are when the modeler chooses to round the data point. We also covered how to create an error table, as well as how to use an error table to decide which model should be selected if there are multiple models to choose from.

©Brian Gillispie, 2016.

1.5: In this Section, we discussed some common area and perimeter models, as well as the distance-rate-time equation. The area and perimeter formulas used were:

Rectangle:
   Area: LW, *Perimeter:* $2L + 2W$

Circle:
   Area: πr^2, *Perimeter:* $2\pi r$

Triangle:
   Area: $0.5bh$, *Perimeter:* $s_1 + s_2 + s_3$

And the distance-rate-time equation was:

$$Time = \frac{Distance\ Traveled}{Speed}$$

Also known as:

$$T = \frac{D}{S}$$

©Brian Gillispie, 2016.

©Brian Gillispie, 2016.

Chapter 2: Data Collection

As we saw in the last Chapter, an important part of mathematical modeling is data collection. In this section, we will discuss how to collect the data, as well as how to set up the data for modeling.

This Chapter will start out with a discussion of the various data types used in modeling. Which data type you use in collecting your data will determine which type of model you can use on the data later on. After we discuss the data types, we will discuss the various techniques used to collect the data.

Once the data is collected, some work needs to be done before any model can be made. First, we need to decide what the independent and dependent variables are. Once those are set, we then need to decide on our numerical scale, as well as what zero means for our model. Finally, we will need to be able to pick a model that has the proper domain and range for the item we wish to model, so we will explore the domain and range of known functions in this Chapter as well.

However, all data collecting needs to follow all the ethical rules involved with data collection. This Chapter will end with a discussion of some of the ethical rules that a modeler needs to keep in mind when collecting data for mathematical model.

©Brian Gillispie, 2016.

©Brian Gillispie, 2016.

2.1: Common Terms in Data Collection

In order to properly model, we need to be aware of the different types of data, as how we represent the data will affect which model we can use on the data. In this Section, we will discuss the common terms used in data collection, as well as how they impact mathematical modeling.

2.1.1: Types of Variables

When modeling, some of the data we collect will be numeric, and some of it will not. For instance, if you are conducting a study to determine which ice cream flavor is the favorite of everyone in the city, the data you collect would most likely not be numeric data. Similarly, if you were to conduct a study to determine how many wolves there are in an ecosystem, the data you collect then would most likely be numeric. For this reason, modelers distinguish between two types of variables, those for which the variable will have numeric answers, and those for which the variable will not have numeric answers. These variables are called quantitative and categorical respectively, and are defined below:

Definition 2.1.1: A variable is **categorical** if the variable can take on one of several categories.

Definition 2.1.2: A variable is **quantitative** if the variable takes on a numerical value.

Notice that these definitions do not say that anything that is reported as a number is automatically quantitative, as you can make numeric categories. Now, let's demonstrate these definitions with an Example or two.

©Brian Gillispie, 2016.

Example 2.1.1: A survey has a blank spot for you to input your age. Is the age variable categorical or quantitative on this survey?

Solution to 2.1.1: Since the input is not forcing you to check a category, you could put in any numeric age you wish. Also, while it is true you could list your age as Bob or Jenny, we will assume that you are going to answer the question seriously here. Therefore, since you will answer the question with a numeric value, this variable is quantitative.

Example 2.1.2: A survey wants you to check a box for your age. The choices are less than 18, between 18 and 35, and over 35. Is the age variable categorical or quantitative on this survey?

Solution to 2.1.2: This time, the survey forces you to choose one of three choices for your age. Even though the categories are numerical, the fact that you are being forced to choose a category means that this variable is categorical.

Notice how in these Examples the same thing was being measured, but in one case the variable was categorical, and in other case the variable was quantitative. Because you can represent the same data as categorical or quantitative data, a modeler needs to decide which one they are going to use before they start collecting the data. The reason is, mathematical operations work really well on quantitative data, as 30 ducks + 30 ducks is 60 ducks. But, mathematical operations do not work as well on categorical data, as many ducks + several ducks is a nonsense operation, even if you intend to use many to represent 30 ducks. Therefore, for this reason, most mathematical modeling uses quantitative variables to collect data, with very few exceptions[26].

---

[26] The exceptions are the month of the year, and the day of the week. We'll discuss how to handle those in a later Section of the book.

©Brian Gillispie, 2016.

2.1.2: Levels of Measurement

All data collected can be classified by the levels of measurement used. To illustrate why this is important, assume you have collected data on worker satisfaction, and the data had every worker rate how satisfied they are with their job on a one to five scale, one meaning very dissatisfied, and five meaning very satisfied. Now, with this scale, does it mean that the worker who votes four is 20% less satisfied than the worker who votes five? And what about the difference from one to two, does it mean a worker who votes two is twice as happy as the worker who votes one, because two is twice one? None of these conclusions are valid conclusions that you could make with this data, as the level of measurement used on the data does not allow you to make these conclusions. For this reason, which level of measurement used is very important in modeling, as it determines what comparisons are valid and what are not valid when we wish to analyze the data.

With that in mind, let's now define the four different levels of measurement used.

> *Definition 2.1.3:* A variable is using **nominal measurement** when the data collected can only be classified and counted. No order exists to the labels.

In other words, a variable was measured nominally if the variable consists of labels for classification. For example, a survey to determine what state each incoming freshmen is from would be collecting nominal data, as the results would be data like Virginia, Arizona, and so on.

> *Definition 2.1.4:* A variable is using **ordinal measurement** when the data can be ordered, but there is no logical interval between the orders.

An example of ordinal data is a survey that asks you to rank your customer service experience as Bad, Poor, Average, Good, or Great. Then, we can safely conclude that a rating of Great is better than Average, but we cannot conclude by how much a rating of Good is better than Average.

©Brian Gillispie, 2016.

> *Definition 2.1.5:* A variable is using **interval measurement** if the data is ordered, and the interval between measurements is a constant size.

At first this sounds exactly like ordinal measurement, but the difference is now, we can now say that two numbers that are 5 units apart means that one number is five better than the other. For example, our temperature scale is an interval measurement. The temperatures 90 and 97 are 7 degrees apart, and so are the temperatures 24 and 31. In both cases, we can safely say those two temperatures are the same distance from each other (7 degrees), but in ordinal measurement, two rankings 7 units apart might not be the same distance from each other, as the rankings are set at arbitrary points.

> *Definition 2.1.6:* A variable is using **ratio measurement** if the data follows the rules for interval measurement, but in addition the unit 0 has meaning, and the ratio between units has meaning.

Ratio measurement looks a lot like interval measurement, but the key difference is that you can now compare ratios between numbers and get a meaningful result. If one worker has a salary of $40000 and another has a salary of $80000, then you would say that one worker is making twice as much as the other. However, if you compare two dress sizes, one a size 6 and one a size 12, it is not true that the size 12 is twice as big as the size 6. That is because with dress sizes, the ratio between two different sizes is meaningless.

So, how does all of this factor into modeling? Since mathematical modeling wants to use the data collected to create an equation that then can be used to make predictions, we are restricted in the types of data collection we can use. For instance, nominal data and ordinal data are pretty useless[27] to us if we wish to create an equation to try and predict real world behavior. Therefore, when modeling, we often use either interval data or ratio data.

---

[27] It is not though totally useless. We can run what is known as a Chi Squared analysis on it still. However, that is considered an advanced technique, and maybe not ideal for what the modeler desires.

©Brian Gillispie, 2016.

2.1.3: Types of Models

In mathematical modeling, we will commonly have to choose between using either a discrete model or a continuous model. Which one is used depends on whether the time variable being used is discrete or continuous, so let's define what is meant by discrete and continuous variables now.

Definition 2.1.7: A variable is **discrete** when the variable can only take on a countable number of values, and there are gaps between the values.

Definition 2.1.8: A variable is **continuous** when the variable can take on an uncountable number of values.

Let's demonstrate these definitions with some Examples now.

Example 2.1.3: You wish to create a traffic model to predict the number of cars on a highway in a given hour. Would the variable cars on the highway be discrete or continuous in this model?

Solution to 2.1.3: In this example, we wish to model the number of cars on a highway. The number of cars can only be zero, one, two, three, and so on. You cannot have 3.5 cars on a highway, so there is a gap between each of those numbers. Therefore, the variable number of cars on the highway is discrete.

Example 2.1.4: You wish to create a model to predict a student's GPA based on how many hours they spent watching television each week. Is the variable student's GPA discrete or continuous?

Solution to 2.1.4: A student's GPA can be any number in the range 0.0 to 4.0, and there are uncountably many numbers in that range, therefore, the variable student's GPA is continuous.

©Brian Gillispie, 2016.

For modelers, the difference between discrete and continuous variables will affect two things. First, whether or not the variable for the result from the model is discrete or continuous will determine if you should round your final answer. For instance, if your model is trying to predict the number of rabbits in an area, and the result is 937.5, since number of rabbits is a discrete variable, you would round that answer to the nearest rabbit. However, if your model is trying to predict a student's GPA, and the result is 3.92, since GPA is continuous, you would report 3.92 as a final answer.

Secondly, whether or not time is discrete or continuous in your model will affect which type of model you use. If you want time to only be measured by specific points in time, then you would use a discrete model. However, if you want time to be able to be measured by any point in time, then you would use a continuous model.

Now, you may wonder why not always use a continuous model for time, since it has the advantage that time can now be measured anywhere? The reason continuous models are not used all the time is they are harder to make and understand[28]. Discrete models, even though they are restricted by only working for those specific points in time, are easier to make and easier to use. Therefore, it is a tradeoff between a model that works everywhere versus a model that is easier to make and use. Which one is used in the end is up to you to decide.

---

[28] Just how complex? Most continuous models require at least a working knowledge of Differential Equations to be able to create them. The ones in this book that are continuous are the few that do not require that knowledge.

©Brian Gillispie, 2016.

2.1 Exercises

Classify the following according to whether the variables are categorical or quantitative. If the question states a blank is given, assume you answer the question properly.

1. A text input on a computer asks for your age.

2. A survey asks you to list your favorite ice cream flavor.

3. A survey wants you to check which political party you support; the choices are Republican, Democrat, or Other.

4. A survey asks for your gender.

5. A graduate school application asks you to write down your college GPA.

6. A W-4 form has a box for you to enter how many dependents you wish to claim on your paycheck.

7. A survey asks you check what day of the week it is. The choices are Monday, Tuesday, Wednesday, Thursday, Friday, Saturday and Sunday.

8. In your own words, state why it is important to know if a variable is categorical or quantitative. How does this affect modeling?

9. In your own words, state why it is important to know which measurement type is used for data collecting. How does this affect modeling?

©Brian Gillispie, 2016.

For Exercise 10 onwards, state the final answer appropriately, depending on whether the final result is discrete or continuous.

10. A model is asked to predict how many cars will be on the road at 8 am, and the model returns 1256.92. How many cars are on the road at 8 am?

11. A model is asked to predict how many people will be in a store at noon, and the model returns 2,562.12. How many people will be in the building at noon?

12. A model is asked to predict the temperature at 3 pm, and the model returns 72.8 degrees. What will the temperature be at 3 pm? *Hint: Assume temperature is continuous.*

13. A model is asked to predict how much snow we will receive from a storm, and the model returns 12.67 inches. How much snow will we receive? *Hint: Assume snowfall totals are continuous.*

14. A model is asked to predict how many children will be enrolled in a school the upcoming term, and the model returns 326.1 children. How many children will be enrolled in the school the upcoming term?

2.2: Independent and Dependent Variables

At some point during the creation of a model, the modeler will need to decide on which data will be input into the model, and which data will be output from the model. In order to do this, the modeler needs to decide on which data point or points are independent variables, and which data point or points are dependent variables. In this Section, we will discuss standards that are commonly used in deciding which data point or points are independent variables and which are dependent variables.

2.2.1: Definitions

To start out, let's list the definitions of each of the terms. The definitions are listed below.

Definition 2.2.1: The **independent variable** (often called the input) is the variable or variables which you have control over, or can manipulate. Oftentimes, the independent variable is the one you think predicts the dependent variable, and are often noted by x in equations or models.

Definition 2.2.2: The **dependent variable** (often called the output) is the variable or variables which you do not have control over, or cannot manipulate. In modeling, the dependent variable (or variables) is the variable(s) we want to predict the value of, and are often noted by y in equations or models.

Based on these definitions, the standard for deciding if a variable is independent or not is to ask yourself: "Does this variable help me to predict another variable?". If the answer to that is yes, then chances are it is the independent variable. Similarly, you can decide if a variable is dependent or not by asking "Is this the variable I want to predict?".

Let's now demonstrate this with a few Examples.

Example 2.2.1: In a study of a book printing company, it was found that if 10 books were printed, it cost the company $250, and if 20 books were printed, it cost the company $350. If we wish to create a model to predict the cost of printing books, based on the number of books we print, which would be our independent and dependent variables?

Solution to 2.2.1: For this problem, we have two variables, cost and books printed. Since we wish to predict the cost of printing the books, based on the number of books we print, then cost is the dependent variable, or y, in this problem. Then, that would make the number of books printed the independent variable, or x, in this problem.

Example 2.2.2: A baseball is tossed into the air. After 1 second has passed, the baseball is 16 feet in the air, after 2 seconds have passed, the baseball is also 16 feet in the air, and after 3 seconds have passed, the baseball is resting on the ground. If we wish to create a model to predict the height of the baseball, based on the time elapsed after tossing the baseball, which would be our independent variable and our dependent variable?

Solution to 2.2.2: For this problem, we wish to predict the height of the baseball, so height will be our dependent variable, or y. Since we wish to predict the height based on the time that elapsed, then time would be our independent variable, or x.

In this book we will often refer to the independent and dependent variable as input and output of the model respectively.

However, be advised that there are exceptions to this rule. Sometimes, people decide to not follow this rule for various reasons. Therefore, you could run into a model where the modeler decided that the input was y and the output was x. This is rare, but it can occur, so just be aware it can happen. Also, note that when time is involved, time is always x in these problems[29], regardless of whether or not we wish to predict time! Let's demonstrate.

---

[29] Though some use t for time instead of x.

Example 2.2.3: A baseball is tossed into the air. After 1 second has passed, the baseball is 16 feet in the air, after 2 seconds have passed, the baseball is 12 feet in the air. Create a model to predict when the baseball will hit the ground, if hitting the ground is represented by a height of 0 feet.

Solution to 2.2.3: The problem states to predict when the baseball hits the ground. Therefore, the independent variable would be distance from the ground, and the dependent variable would be the time since the baseball was released. However, since time is involved, we would let x be time, since x is always time when time is in a problem. Therefore, y is now distance from the ground, even though that flips the standard for independent and dependent variables!

That last Example might have left you confused, as now we have an exception to our rule. The easiest way to remember it is this: Follow the rules for independent and dependent variables **except** when time is involved. If time is involved, time is x, no matter what.

Now, the reader might also be wondering why it matters if we set our input and output variables correctly. After all, in many models, there is just an x and a y, and if they are flipped around, as long as we remember what the variables mean we are still fine. The reason it matters still is twofold. One, we want to set our inputs and outputs correctly for consistency with other literature. All science and biology books follow this standard for independent and dependent variables, so in an attempt to remain consistent we will follow the same standard. The second reason is because you cannot always get away with flipping the variables, despite what you may think. For example, assume the model we created to predict the distance a particle has traveled after x seconds is given by the following:

$$y = -16x^3 + 10x^2 - 33.33x + 12$$

Now, if you wish to use this to predict what the distance is after 5 seconds elapsed, you would plug in 5 for x, and solve the following:

$$y = -16(5)^3 + 10(5)^2 - 33.33(5) + 12$$

©Brian Gillispie, 2016.

This is a simple calculation with a calculator. However, if you reversed the meaning of x and y in the model, now in order to predict the distance traveled after 5 seconds have elapsed, you would need to solve the following equation[30]:

$$5 = -16x^3 + 10x^2 - 33.33x + 12$$

Which is hard, if not impossible, to solve.

2.2.2: Selecting Variables from Problems

Now that the standard is set for how to select your input and output for your model, we are going to work with a few math problems to demonstrate how to properly select the points for creating the model. Many of these problems will be returned to later in the book, so feel free to return to this Section when you are unsure how points are being selected.

To begin, when it comes to mathematics problems, start by determining what it is you wish to predict. Is it the number of rabbits? The speed of sound? The height of a baseball? Then, determine what it is you can use to predict this value. Once you have that set, decide on what the variables are, based on the rules of the Section, then assign all associated values together as points, being sure to keep all associated values together in the same point. Let's demonstrate this now with some Examples.

Example 2.2.4: A small town is slowly losing its population as people move away to the big city. In the year 1990 the town had 675 people, and in the year 1991 the town had 663 people. Predict how many people will be in the town in 1992, based on this data.

Solution to 2.2.3: In this Example, the entire problem is stated, but we are only interested in how the points are selected at this point in time. To do that, first we have to determine what we wish to predict. In this problem we wish to predict the population of the town in a future year, so the

---

[30] We should point out that the equation would not be exactly the same due to the flipped variables, but you would still have a cubic equation, which cannot be solved except in very special cases.

©Brian Gillispie, 2016.

population of the town should be our dependent variable, or y. Then, since the time is what determines the population of the town, time is our independent variable, or x.

With both of the variables determined, now we look at the problem and see what information we know. We know that in 1990, there were 675 people. Since those two pieces of information are associated with each other, we will represent the data by the point $(1990, 675)$. Notice that is a (x, y) point, with x listed first and y listed second. What else do we know? We know that in the year 1991, there were 663 people, which can then be represented by the point $(1991, 663)$.

Therefore, the two points that we have, based on this information, are $(1990, 675)$, and $(1991, 663)$, with x representing the year and y representing the population of the town.

Example 2.2.4: A farmer starts out with 10 chickens. One month later, he has 12 chickens, and two months later he has 14 chickens. Predict how many chickens he will have five months later, if this trend continues.

Solution to 2.2.4: Again, we are only interested in determining what points we have to work with to create our model, based on this information. As such, a complete solution will not be presented.

First, start out by deciding what the independent and dependent variables are. In this problem, we wish to predict the number of chickens in the future, so the number of chickens will be our dependent variable, or y. Also, since we wish to predict the number of chickens based on the current month, then the current month will be the independent variable, or x.

With the variables decided on, look back on the problem and see what data has been provided. We know that after one month has passed, the farmer has 12 chickens. This can be represented by the point $(1, 12)$. Also, we know that after two months have passed, the farmer has 14 chickens. This can be represented by the point $(2, 14)$.

However, this is not all the information available to us this time. Recall that the farmer started with 10 chickens. This can be represented as a point too, as starting with means zero time has passed. Therefore, this can be represented by the point $(0, 10)$.

Therefore, we have three points to work with, based on this information, $(0, 10)$, $(1, 12)$, and $(2, 14)$.

©Brian Gillispie, 2016.

Sometimes in modeling, the information is presented in table format instead. If the data is in table format, the same standards apply to picking out points. However, if no information is given in the question to allow one to determine which should be the independent and dependent variable, then the modeler has to just choose. In this book, we will always use x for the values in the first column of a table, and y for the values in the second column of a table, unless told otherwise. So for example, if we had the following table:

Month	Number of Ducks
0	25
1	44
2	52

Then, since we are not given any information about what the model is supposed to predict[31], we would assume that the value in the first column (months) is then to be represented by x, and the value in the second column (number of ducks) is to be represented by y. This would mean that our data points would then be $(0, 25)$, $(1, 44)$, and $(2, 52)$.

Going forward in this book, we will use the standards in this Section to determine what the variables x and y are to represent, as well as to select our data points for problems.

---

[31] This is rare in practice, but it can happen. When this does happen, usually the modeler has to decide what they wish to predict, then base the variables on that decision.

©Brian Gillispie, 2016.

2.2 Exercises

For all the Exercises, state the data points in (x, y) format, and define x and y. Do not solve, as you will later be asked to solve these problems using the appropriate models in Chapters 3 and onwards.

1. A pond has 44 ducks in the year 2000, and 56 ducks in the year 2010.

2. At 1 pm it was 68 degrees outside, and at 2 pm it was 70 degrees outside.

3. At 10 am there were 67 pigeons on the telephone wire, and at 11 am there were 69 pigeons on the telephone wire.

4. A pack of wolves has 25 wolves in the year 2000, and 26 wolves in the year 2001.

5. At 5 pm a culture of bacteria has 1000 cells, and at 6 pm a culture of bacteria has 2000 cells.

6. A town has 2500 people in the year 2015, and 2566 people in the year 2016.

7. A pack of wolves has 25 wolves, and 4 years later the pack has 30 wolves.

8. A lake has 10 grams of a pollutant, and 3 months later, the lake has 9 grams of the pollutant.

9. A ball is tossed into the air. When the ball is released, it is zero feet in the air. After a second has passed, the ball is 72 feet in the air.

10. There are currently 1500 books in storage, and next year estimates are there will be 1600 books in storage.

©Brian Gillispie, 2016.

2.2: Independent and Dependent Variables 84

11. A river has 10000 salmon, and next year the river will have 11000 salmon.

12. Currently, 3 inches of snow has fallen. An hour from now, 4 inches of snow will have fallen.

13. The river is 12 feet high today, and will be 12.5 feet high tomorrow.

14. A farmer has 25 chickens initially, and will have 27 chickens one month from now and 29 chickens two months from now.

15. A ball is 4 feet in the air initially, 10 feet in the air after one second has passed, and 2 feet in the air after two seconds have passed.

16. A town has 6000 people in the year 2020, 5750 people in the year 2021, and 5600 people in the year 2022.

17. On a remote island, there are 32 rabbits in the year 2020, 675 rabbits in the year 2021, 4,780 rabbits in the year 2022, and 23,417 rabbits in the year 2023.

©Brian Gillispie, 2016.

2.3: Numerical Scales

After deciding on what the independent variables and dependent variables will be in a model, a modeler has to decide on which numerical scale they wish to use for the variables in the model. Do they want inputs for the model to be in ones or hundreds? Do they want input to be in years after 2000, or years after 2010? All of these will affect what form the points in the data set will take on, which will in turn affect the model. Therefore, it is important to decide in advance what scale the model will be using.

In this Section, we will discuss both how to use a scaled model, as well as how to set a scale for a model.

2.3.1: Using a Scaled Model

Most mathematical models are set up so that some sort of number scale is used in the model. For that reason, you cannot assume that you can just plug in the number given into the model to get the desired answer, as sometimes you need to rescale your number before you plug it in. Let's demonstrate.

Example 2.3.1: A model to predict college tuition for the school year is of the form $y = 4000 * 1.03^x$, where x is in years after 2000. Predict how much college tuition will be in the year 2005 with this model.

Solution to 2.3.1: Be careful. A common error is to plug in 2005 for x, which would then claim that the tuition in 2005 is the following:

$$y = 4000 * 1.05^{2005} = 1.22 * 10^{46}$$

That is a **lot** of money for college tuition. In fact, that is more than the current US national debt, and is probably more than the sum of all money in the entire world, combined.

©Brian Gillispie, 2016.

Obviously, that answer is unreasonable. The reason though we got an unreasonable answer was we blindly plugged in 2005 and didn't read how the variable x was set up in the model. In this model, x was set up to mean years after 2000, so when you plug in for x, you need to plug in how many years after 2000 the given year of 2005 is, which is found by computing $2005 - 2000$. Therefore, the correct answer to this problem is:

$$y = 4000 * 1.05^5 = 5015.13$$

Or, college tuition will be approximately $5015.13 in the year 2005, based on this model.

Example 2.3.2: A model to predict your weight based on your height is given as $y = 3.6x + 120$, with x your height in inches over 5 feet. Predict how much someone who is 6 feet tall would weigh in this model.

Solution to 2.3.2: This time there are two things to watch out for. First of all, you wish to predict how much someone weights based on their height, but the model wants the height in inches, and you have height in feet. Therefore, we need to convert the height to inches first[32]. Doing that conversion tells us now that we wish to predict how much someone will weigh if they are 72 inches tall.

Now, we will still need to be careful, as the model specifies that the input must be in inches over 5 feet, or 60 inches, as 5 feet is 60 inches. To find out how many inches we are over 60 inches, we compute $72 - 60$, which gives us 12. Then, we plug 12 into the model, which gives us that this person weighs:

$$y = 3.6(12) + 120 = 163.2$$

Therefore, based on this model, we would predict that someone who is 72 inches (or 6 feet) tall would weigh 163.2 pounds.

---

[32] This is **always** the rule for word problems. Put every single thing related to the input in the same units. No exceptions.

©Brian Gillispie, 2016.

Always check your model carefully before plugging in the points. The most common error that is made when using someone else's model is the user didn't pay careful attention to the units the model was expecting, and therefore plugged in 225 pounds when the model wanted weight in kilograms, or something similar.

2.3.2: Data Point Scaling

Now that we have demonstrated how to use a previously scaled model, let's discuss how we can scale our data points, which is where we convert our data points to either new units or a new scale (or both). Most times this conversion requires a quick multiplication or division to change to the new units, but sometimes multiplication or division is not appropriate, as you are only changing where in the number scale your points are at. Let's at this point in time demonstrate a couple of cases where we change to new units, and a couple of cases where we shift where in the number scale our points are at[33].

Example 2.3.3: A researcher has collected the following data points, with x in weeks, and y in frogs:

$$(0, 100), (1, 110), (2, 125), (3, 140)$$

Convert the data points so that x is in days instead.

Solution to 2.3.3: Since there are 7 days in a week, this requires us to multiply the x value by 7 in each data point to covert the values to weeks. Therefore, once we do that, the new points are:

$$(0, 100), (7, 110), (14, 125), (21, 140)$$

Notice the y values on each point are untouched, as they were not being scaled.

---

[33] Caution: This section assumes you know how to convert between different units.

Example 2.3.4: A researcher has collected the following data, with x in hours and y in bacteria cells:

$$(0, 200), (4, 445), (8, 900), (12, 1995), (16, 4125).$$

Convert the data points so that x is in four hour intervals.

Solution to 2.3.4: In this Example, we want to convert the data points so that an input of x is how many four hour intervals have passed. This means that if x is 5, that means 5 4 hour intervals have passed, or 20 hours[34] have passed. Therefore, to do this conversion, since x is in hours, we divide all the x values by 4. Once we do that, the new data points are:

$$(0, 200), (1, 445), (2, 900), (3, 1995), (4, 4125)$$

Example 2.3.5: While recording data on sales at a big business, you collected the following data, with x in years and y in sales (in millions):

$$(2006, 15.6), (2007, 16.2), (2008, 9.8), (2009, 11.7), (2010, 14.1)$$

Convert the data points so that x is in years after 2000 instead.

Solution to 2.3.5: This is an example of an additive shift, where this time we are scaling x so that it is still in years, but each x value will now tell you how many years after 2000 each individual point is. Therefore, in order to do this, we need to subtract 2000 from each of the x values. Once we do that, the new data points are:

$$(6, 15.6), (7, 16.2), (8, 9.8), (9, 11.7), (10, 14.1)$$

---

[34] This is found by taking the number of four hour intervals (5) times the length of each interval (4).

©Brian Gillispie, 2016.

Example 2.3.6: A business has decided to keep count of how many employees it hires ever year. Currently it has the following data points, with x in years and y number of new employees hired:

$$(2013, 122), (2014, 17), (2015, 226), (2016, 109), (2017, 196)$$

Convert the data points so that x is in years since the business as founded, which is the year 1936.

Solution to 2.3.6: This time we need to shift the scale so that x is in years after 1936. Like before, we are not changing units, just changing how the units are displayed. To convert our x values so that each of them displays years after 1936, we need to subtract 1936 from each of our data points. If we do that, we end up with:

$$(77, 122), (78, 17), (79, 226), (80, 109), (81, 196)$$

Before we close this Section, we should discuss something quickly. In two of our Examples we had to multiply or divide to find our new units, and in two of our Examples, we had to add or subtract to find our new data points. When do we know when to do which? The easiest way to remember this is ask yourself if you are changing units or not. If you are changing the units the point is displayed in, like say going from days to weeks, then you would multiply or divide. However, if the new data points are in the same units but just at a different point on the scale, then you would add or subtract, as all you are doing now is changing how the points are displayed.

2.3 Exercises

For Exercises 1 - 9, evaluate the model for the given point.

1. $y = 1.2x + 55$, x in hours, $x = 30$ minutes

2. $y = 120 * 0.65^x$, x in days, $x = 48$ hours

3. $y = 2.5x + 1.65$, x in years after 2000. $x = 2016$

4. $y = 10.56 * 1.04^x$, x in years after 2007, $x = 2015$

5. A model to predict the number of geese in a forest is given by the following: $y = 55 * 1.1^x$, with x in months. Predict how many geese will be present 75 days from now? *Assume 30 days in a month.*

6. A model to predict how much radiation is still present after a scan is given by $y = 53 * 0.8^x$, x in hours after the scan. Predict how much radiation will be present 3 days after the scan.

7. A model to predict how much snow has fallen during a snowstorm is given by $y = 1.14x + 2$, with x in hours and y in inches. Predict how much snow has fallen 180 minutes later.

8. A model to predict how much rain has fallen during a really heavy rainstorm is given by $y = 0.08x + 0.2$, with x in minutes and y in inches. Predict how much rain has fallen 360 seconds later.

9. A model to predict how many calories you burn during a walk is given by $y = 200x$, with x in miles and y in calories. Predict how many calories you will burn during a 15 foot walk with this model.

©Brian Gillispie, 2016.

2.3: Numerical Scales

For Exercises 10 onwards, scale the given data points as directed.

10. A researcher has collected the following data points, with x in minutes, and y the amount of medicine in the patient: $(0, 325)$, $(60, 305)$, $(120, 287)$, $(180, 274)$. Shift the x values so that they are in hours instead.

11. For the data points in Exercise 10, shift the x values so that they are in 15 minute intervals instead.

12. A business has collected the following data points, with x in year and y in sales in millions of dollars: $(1992, 5.4)$, $(1993, 5.2)$, $(1994, 5.5)$, $(1995, 6.1)$, $(1996, 6.0)$. Shift the x values so that they are in years after 1900 instead.

13. For the data points in Exercise 12, shift the x values so that they are in years after 1975 instead.

14. While in the lab, you collected the following data on the growth of a bacteria, with x in minutes, and y in cells: $(300, 1000)$, $(360, 1075)$, $(420, 1150)$, $(480, 1269)$, $(570, 1501)$. Shift the x values so that they are in hours.

15. For the data points in Exercise 14, shift the x values so that they are 30 minute intervals instead.

16. For the data points in Exercise 14, shift the x values so that they are in minutes after 300 minutes.

17. For the data points in Exercise 14, shift the y values so that they are in thousands of cells instead.

©Brian Gillispie, 2016.

18. A quality assurance representative collects the following data on the number of defective products that the company has returned, with x in years and y the number of defects returned: $(2005, 122), (2006, 178), (2007, 115), (2008, 99), (2010, 350)$. Shift the x values so that they are in years after 2000.

19. For the data points in Exercise 18, shift the x values so that they are in years after 1899 instead.

20. The following data was collected on a pollutant in a lake, with x in days, and y in liters of polluted water: $(8, 12.9), (9, 11.5), (10, 11.6), (11, 10.2), (12, 9.6), (13, 9.9)$. Shift the x values so that they are in hours instead.

21. For the data points in Exercise 20, shift the x values so that they are in 3 day intervals instead.

22. For the data points in Exercise 20, shift the x values so that they are in days after day 8.

23*. For the data points in Exercise 20, shift the x values so that they are in days after day 8, and in 2 day intervals.

24. For the data points in Exercise 20, shift the x values so that they are in days after day 5, and the y values so that they are in milliliters.

2.4: Handling Categorical Data

While collecting data, sometimes a modeler will be forced to use categorical data when recording data points. In this section, we discuss how one handles those data points when they arise, and how this impacts using those points when creating a mathematical model.

2.4.1: Converting Categorical Data to Quantitative Data

In order to use categorical data in a mathematical model, the data needs to be converted into a quantitative format. The reason for this is if you don't, you may suddenly be required to calculate 6 * Tuesday + Friday, which has no meaning. However, categorical data by themselves have no numbers. Therefore, it is up to the modeler to assign numbers to the categorical data so that mathematical operations have meaning again.

Which numbers to assign to categorical data is often the subject of long debates, as the number assignment is arbitrary. The only restriction is once you assign a number to a category, you can't use that number again unless it means the same category. This means that if you decide that Tuesday is the number 4, you can only use the number 4 to represent Tuesday going forward. Otherwise, any number assignment is valid.

Let's now demonstrate this with a few Examples before moving on:

Example 2.4.1: While collecting data for sales for a store, you recorded the data points in the format (Day of Week, Sales), and the data points are as follows: (Tuesday, $2669), (Wednesday, $4015), (Thursday, $1980), (Friday, $2890). Report the data points with the x value in quantitative form.

Solution to 2.4.1: Answers to this will vary, as the choice of what number to assign to each day is arbitrary. That means that one could assign Tuesday to 0, Wednesday to 1, Thursday to 5 and Friday to -10, which would convert the data points to:

$$(0, 2669), (1, 4015), (5, 1980), (-10, 2890)$$

However, since the choice of number is arbitrary, this is also equally valid:

$$(2, 2669), (3, 4015), (4, 1980), (5, 2890)$$

As is this:

$$(5.62, 2669), (1090, 4015), (-999999, 1980), (3.14159, 2890)$$

All of those assignments are valid, as none of them assign two different categories to the same number.

Example 2.4.2: While observing ducks on a pond, you recorded the data points in the form (Months, Ducks), and the data points are as follows: (May, 45), (June, 54), (July, 81), (August, 85), (September, 76). Report these data points with the x value in quantitative form.

Solution to 2.4.2: Answers will vary like they did in Example 2.4.1, as the only restriction is you cannot assign two months to the same number. So this means we could assign May to 2, June to 5, July to 7, August to 8 and September to 14, which would make the data points:

$$(2, 45), (5, 54), (7, 81), (8, 85), (14, 76)$$

This is a valid assignment (though a poor one); if no other restrictions are given.

At this point the reader has probably noticed that some assignments work better than others. It is really confusing to work with a model that decides that May is the number 2, as we are used to May being the fifth month of the year! Also, the assignment in Example 2.4.2 has another, less obvious problem, and that is if we take 2 and add 1 to it, we get 3. However, the way we assigned our values, the next month after May is June, which means for our data points to work well in a model, we need $2 + 1 = 5$ now. Otherwise, if we ever increase our x value by one, it will not move our input over to the next month like we expect it to! Therefore, in order to make this happen, we need to add another restriction to how we assign numbers to quantitative variables. This restriction is that when we increase our assigned number by 1, we should move into the next ordered category in line. Or, in other words, whatever number we assign to May, we need to assign to June the next number in line, as June is after May. If you do this, all number assignments will then be consistent in the model, and mathematical operations will work as expected as well.

In summary, the rules for assigning numbers to categorical data can be summarized as follows:

> ***Rules for Assigning Numbers to Quantitative Data:***
>
> **1**: No two categories can be assigned to the same number.
> **2**: If you add 1 to the number assigned to one category, you end up with the number assigned to the next category.

What this means is once you have decided on what number to assign to the first category, the rest will follow suit quickly. Let's demonstrate.

Example 2.4.3: While collecting data on the population of a town, you recorded the points in the form (Month, People), and your data points are as follows: (May, 4001), (June, 2879), (July, 2901), (August, 3506), (September, 4167). Report these data points with the x value in quantitative form.

Solution to 2.4.3: Answers will vary, but one solution is to assign to May the number 5. Then, since adding 1 to 5 gets 6, and June is after May, June has to be assigned to the number 6. Similar reasoning gets that July has to be assigned to the number 7, and so on. Continuing like this gives us the following data points:

$$(5, 4001), (6, 2879), (7, 2901), (8, 3506), (9, 4167)$$

©Brian Gillispie, 2016.

Example 2.4.4: While collecting data on how many hamburgers your business sells, you record the data in the form (Day, Burgers), and your data points are as follows: (Wednesday, 45), (Thursday, 22), (Friday, 56), (Saturday, 71). Report these data points with the x value in quantitative form.

Solution to 2.4.4: Answers vary again, as long as you are consistent. One solution is to say that Wednesday is 0. Then, since $0 + 1$ is 1, and Thursday is the day after Wednesday, then Thursday has to be 1 if Wednesday is 0. Continuing like this for all the data points then will give us the following:

$$(0, 45), (1, 22), (2, 56), (3, 71)$$

Now, someone is probably wondering why didn't I use 3 for Wednesday, since Wednesday is the third day of the week? The reason I didn't is I don't have to! The choice of number to assign to the day Wednesday is completely arbitrary, which means you can assign to it any number you wish, as long as the number one more than that is assigned to Thursday as well.

2.4.2: Some Issues

When a modeler uses categorical data and assigns numbers to them so they can use mathematical equations on the data set, there are some issues they need to be aware of. First, the modeler needs be careful that they only attempt to apply these rules to data that is either ordinal, interval, or ratio data. The reason is if you attempt to apply these rules to nominal data, then you can end up with a situation where Amy is 1, Jeff is 2 and Brian is 3, so Amy + Jeff is Brian because $1 + 2 = 3$. Or, even crazier, if you decide to use 1 for Male, 2 for Female, then what is $1 + 2$ now? Since all of your number assignments can be added to (and more) you have to account for what those results will mean as well. However, since nominal data doesn't even have clear intervals in the data sets, this is impossible when working with nominal data.

Even if the modeler is working with data that is not nominal, there are a couple of other issues to be aware of, and that is how to handle data assignments that are on the boundary. Let's demonstrate this with an Example.

Example 2.4.5: A modeler has recorded data on sales for a store, and recorded the following data, in the form (Date, Sales): (May 30th, $2,669), (May 31st, $2,891), (June 1st, $2,481), (June 2nd, $2,656). Report these data points with the x value in quantitative form.

Solution to 2.4.5: We have to be a little careful here, because the temptation is to assign May 30th to 30, May 31st to 31, and June 1st to 1. However, the rule requires that if we add 1 to May 31st's assigned number, we get June 1st assigned number, so June 1st would have to be assigned 32 now, as 31 + 1 is 32. Likewise, this means that June 2nd is now assigned 32 + 1, or 33. Therefore, our data points are now as follows:

$$(30, \$2{,}669), (31, \$2{,}891), (32, \$2{,}481), (33, \$2{,}656)$$

Notice what happened here. Because of the rule that required that whatever number we assigned to the next date was 1 more than the previous date's number, we could not use 1 for June 1st, despite the fact that is what most people want to use. Always check for if you are following the rule that adding 1 to the previous number gets you the next number. This may result in you having to assign Monday to 8 instead of 1 due to Sunday being assigned 7, or having to assign January 2016 to 13 because you assigned December 2015 to 12. Always check to see if your assignment follows the rules given earlier for number assignment for categorical data, otherwise your model will not perform as expected and you will end up with very unusual results in the end.

©Brian Gillispie, 2016.

2.4 Exercises

For the following problems, one or both of the data points will be in categorical form. Convert the data points to quantitative, and state what each of your assigned numbers to the categories mean. Make sure all assignments follow the rules on Page 95 unless the data is nominal.

1. Points are in the form (Day of Week, Customers):

    $(Monday, 72), (Tuesday, 55), (Wednesday, 60), (Thursday, 102), (Friday, 98)$

2. Points are in the form (Gender, Height):

    $(Male, 72), (Female, 68), (Male, 66), (Female, 62), (Male, 70), (Female, 71)$

3. Points are in the form (Date, Number of infected):

    $(Jan\ 12th, 150000), (Jan\ 13th, 167900), (Jan\ 14th, 177090). (Jan\ 15th, 172950)$

4. Points are in the form (Month, Owls):

    $(March, 25), (April, 32), (May, 37), (June, 42), (July, 55)$

5. Points are in the form (Date, Ice Cream Flavor):

    $(July\ 21st, Vanilla), (July\ 22nd, Chocolate), (July\ 23rd, Strawberry)$

6. Points are in the form (Game Number, Football Team):

    $(1, Bears), (2, Lions), (3, Bengals)$

7. Points are in the form (Show, average ticket price):

    $(Cinderalla, \$50), (Beauty\ and\ the\ Beast, \$55), (RENT, \$48)$

8. Points are in the form (Name, Age):

    $(Kate, 22), (Kurt, 23), (Alyssa, 21), (Aaron, 24), (Julia, 22), (John, 26)$

©Brian Gillispie, 2016.

9. Points are in the form (Month, Number of missed days):

$$(Dec\ 2014, 3), (Jan\ 2015, 7), (Feb\ 2015, 1), (March\ 2015, 0)$$

10. Points are in the form (Day of Week, Sales). Days are in sequential order.

$$(Friday, 12), (Saturday, 5), (Sunday, 0), (Monday, 0), (Tuesday, 17)$$

11. Points are in the form (Date, customers):

$$(June\ 29th, 325), (June\ 30th, 299), (July\ 1st, 311), (July\ 2nd, 349)$$

12. Points are in the form (Week, Customers):

$$(August\ 9-15, 3415), (August\ 16-22, 5671), (August\ 23-29, 1990)$$

13. Points are in the form (Month, Ducks):

$$(Oct\ 2016, 14), (Nov\ 2016, 7), (Dec\ 2016, 0), (Jan\ 2017, 0), (Feb\ 2017, 2)$$

14. Points are in the form (Month, Birds). Some months are missing.

$$(Oct\ 2018, 908), (Nov\ 2018, 325), (Mar\ 2019, 25), (Apr\ 2019, 266), (July\ 2019, 2090)$$

15. Points are in the form (Month, Average High Temperature):

$$(Mar\ 2015, 56), (June\ 2015, 90), (Sept\ 2015, 69), (Dec\ 2015, 24), (Mar\ 2016, 48)$$

2.5: Domain and Range

When working with models, it is important to know the domain and range of the model being used. Otherwise, you might get nonsense results like 200% of the United States smoked in the year 1860, or there are $1.45 * 10^{57}$ rabbits in Australia now. In this Section, we will discuss what is meant by domain and range, and then will discuss how to work with domain restricted models.

2.5.1: Domain

We will start our discussion by defining the domain of an equation. The definition is as follows:

Definition 2.5.1: The **domain** of an equation is the set of valid inputs for the equation

In other words, the domain of a mathematical equation is the set of inputs that one can plug into the equation and receive a valid answer. Usually with equations, the domain of the equation is all real numbers **except:**

1: Any input that causes a division by zero to occur.
2: Any input that causes the equation to have to compute a negative square root.
3: Any input specifically restricted from the domain of the equation.

Case three is special and will be discussed a little later. For now, let's consider a few Examples of equations where we need to consider cases one and two first.

©Brian Gillispie, 2016.

Example 2.5.1: Find the domain of the equation $y = 2x + 7$

Solution to 2.5.1: The equation has no division steps, so there is no possibility of division by zero occurring. Similarly, the equation has no square roots in the function, so it is not possible for any negative square roots to be taken. Therefore, the domain of this equation is all real numbers, often denoted as $(-\infty, \infty)$

Example 2.5.2: Find the domain of the equation $y = \frac{2}{x-3}$

Solution to 2.5.2: In this equation, there are no square roots, but there is a division step. Since we cannot divide by zero, for that reason, $x - 3$ cannot equal zero. However, if $x = 3$, then $x - 3$ will equal zero. Because of that, the point 3 needs to be removed from the domain of the equation, which means the domain of the equation is all real numbers except for when x is 3. This is often denoted in interval notation as $(-\infty, 3) \cup (3, \infty)$.

Example 2.5.3: Find the domain of the equation $f(x) = \sqrt{x-4}$

Solution to 2.5.3: This time there is a square root, but no division step. We cannot compute the square root of a negative number[35], so this means that $x - 4 \geq 0$ must be true for the equation to produce a valid output. For $x - 4 \geq 0$ to be true, then $x \geq 4$ must be true as well.

Therefore, the domain for the equation is all x, such that $x \geq 4$.

In these first few Examples, we assumed the domain had no prespecified conditions. Usually this is the case; however, sometimes we do wish to intentionally restrict the domain of our model. The reason for this is mathematical models will take **any** input you provide, and return an answer for that input. So if you decide to input the value 365 days into a snowfall model predicting snowfall from a storm, you might get the result that after 365 days have passed, the total snowfall from the storm is 1,920 inches. However, in reality, the storm has long since ended (hopefully), and that result is utter nonsense. In an attempt to prevent these kinds of

---

[35] That is, unless we wish to allow for complex numbers in our model, which we are not allowing.

©Brian Gillispie, 2016.

nonsense answers resulting from a model, a domain restriction is applied, with the restriction intending to denote in what range of inputs the model is still useful. An example of a domain restriction in an equation is as follows:

$$y = 2x + 7, \quad 0 \leq x \leq 10$$

Notice the inclusion of the $0 \leq x \leq 10$ right after the equation? That is telling the person using the model that there is a restriction on the variable x, which is that x must be between the numbers 0 through 10. And, since x is the input to this equation, this restriction has to be applied to the domain as well.

What does this mean to us when we wish to use a model with a domain restriction? It means that we can only plug into the model inputs that are inside the specified domain. If the input is not in the specified domain, the model will return an error. What the error looks like will vary, but in this book we will say the answer does not exist (DNE) when an input outside the domain is plugged into the model. Let's now demonstrate this with a few Examples.

Example 2.5.4: Given the equation $y = 10 * 2^x$, $0 \leq x \leq 5$, evaluate the equation for $x = 4$

Solution to 2.5.4: Since an x value of 4 is inside the restriction applied to x, we can plug it in and solve as normal. Doing that gives us:

$$y = 10 * 2^4 = 160$$

Therefore, when x is 4, y is 160.

Notice that the fact that y was outside the range does not matter, as the restriction is only applied to x. Be very careful[36] to only apply the condition to x when working these problems.

---

[36] And I mean be careful! The most common error when working these problems is to accidentally apply the condition to y, not x.

©Brian Gillispie, 2016.

Example 2.5.5: Given the equation $y = 10 * 2^x$, $0 \leq x \leq 5$, evaluate the equation for $x = 6$

Solution to 2.5.5: This is the same equation as in Example 2.5.4, but this time our given input of 6 is outside the given range for x. Therefore, the model is not set up to handle that input, and the answer is DNE.

Example 2.5.6: Given the equation $y = -16x^2 + 64x$, $0 \leq x \leq 4$, evaluate the equation for $x = 1$.

Solution to 2.5.6: The input of 1 for x is inside the range given, so we can plug it in and compute as normal. Doing this gives us:

$$y = -16(1)^2 + 64(1) = 48$$

Therefore, when x is 1, y is 48.

Example 2.5.7: Given the equation $y = -5x + 75$, $0 \leq x \leq 17$, evaluate the equation for $x = 25$.

Solution to 2.5.7: The input of $x = 25$ is outside the range of $0 \leq x \leq 17$, so therefore the model is not set up to handle that input, and the answer is DNE.

Currently, we have said nothing on how these domain restrictions are found and set up on a model. After all, why is the domain for Example 2.5.7 $[0, 17]$, and not say $[-10, 25]$? Domain restrictions are usually set up in a way such that the values that are still accepted are reasonable for the model given. So for Example 2.5.7, maybe the creator of the model thought an output of a negative number was unreasonable, so they cut off the model at an input of 17? Unfortunately, the decision of where to apply the domain restriction is slightly arbitrary, and logic and deduction is needed in order to set these up decently. Usually this logic requires being able to find when the model hits certain values for the outputs, so we will say no more on this until we have explored a few mathematical models in more detail.

©Brian Gillispie, 2016.

2.5.2: Range

The range of a function is defined as follows:

> *Definition 2.5.2:* The **range** of the function f(x) is the set of all valid outputs of the function.

The range of a function tends to be harder to find than the domain. We can usually come up with an instinctive idea of what kind of outputs that an equation or model will produce, but how do we prove that this is all of the possible outputs? How do we know that there is not some oddball input that will produce an output we did not expect? For this reason, range is difficult to find and justify algebraically. However, sketching a graph of the function is a good way to estimate the range, though the graph is, by itself, not enough justification that the range is what you claim it is. Usually, the range is found through a mathematical proof, and justified that way. For the interested reader we included one proof of range at the end of this Section.

Since justifying the range of the function is non-trivial, we will only focus on learning the range of functions where the range has already been proven by others.

2.5.3: Domain and Range of Known Functions

Before proceeding, we wish to now discuss some known functions and their domain and ranges. Learning this information will be important for when we decide to model with these functions later, because the modeler needs to be aware of what inputs the model can receive if they do not apply any restrictions to the inputs[37]. Please refer back to this Section as needed once we discuss the appropriate models later in the book.

The first function to discuss the domain and range for is the general equation of the line $f(x) = mx + b$, which we will learn more about in Chapter 5. The domain and range is as follows:

---

[37] Unfortunately, you cannot restrict the outputs, short of removing the input that would cause that output from the domain of the function.

©Brian Gillispie, 2016.

> Given the function $f(x) = mx + b$, the domain and range of the function is as follows:
>
> Domain: $(-\infty, \infty)$
>
> Range: $(-\infty, \infty)$

This means then that for the function $f(x) = mx + b$, the domain is any real number and the range is any real number. This is true for any line, provided that $m \neq 0$ (see if you can figure out why m cannot be zero for this condition to be true).

Next up, we will discuss the domain and range for the function $f(x) = a^x$, where $a > 0$. This is a general exponential function, and is discussed more in Chapter 6. The domain and range for the general exponential function is as follows:

> Given the function $f(x) = a^x$, with $a > 0$, then the domain and range of the function are as follows, provided $a \neq 1$:
>
> Domain: $(-\infty, \infty)$
>
> Range: $(0, \infty)$

This means that an exponential function can receive any input, but can never return a negative number or a zero as an output. This is important to remember in modeling, because if one of your outputs needs to be zero, then an exponential model is a poor choice for a model.

Next, we will discuss the domain and range for the general quadratic function, which we will use in Chapter 7. For the general quadratic $f(x) = ax^2 + bx + c$, it can be shown that the domain and range of the function is as follows:

©Brian Gillispie, 2016.

> Given the function $f(x) = ax^2 + bx + c$, and $a \neq 0$, then the domain and range of the function are as follows:
>
> Domain: $(-\infty, \infty)$
>
> Range: If $a > 0$, then the range is $[d, \infty)$, if $a < 0$, then the range is $(-\infty, d]$, where $d = f(\frac{-b}{2a})$

This one is a little more complicated, due to the fact that a quadratic function does not return every possible output. In Chapter 7 we will discuss how to find that number d mentioned above in more detail. For now, just be aware that a quadratic either has a maximum or a minimum, but not both.

The next function we wish to discuss the domain and range of is the trigonometric function $f(x) = a * sin(x) + b$, which we will use in Chapter 7 as well. It can be shown that the domain and range of this trigonometric function is as follows:

> Given the function $f(x) = a * sin(x) + d$, with $a \neq 0$, the domain and range of the function are as follows:
>
> Domain: $(-\infty, \infty)$
>
> Range: $[-|a| + d, |a| + d]$

This means that the trigonometric function $a * sin(x) + d$ can receive any input, but the output covers a small range.

Finally, the last function we will discuss the domain and range of in this section is the logarithmic function $f(x) = \ln(x)$, which we will use some in Chapter 6 and 7. It can be shown that the domain and range of the ln function is as follows:

©Brian Gillispie, 2016.

> Given the function $f(x) = \ln(x)$, the domain and range of the function are as follows:
>
> Domain: $(0, \infty)$
>
> Range: $(-\infty, \infty)$

Notice that the value of zero is not included in the domain of $\ln(x)$. However, the range is all real numbers.

Now that the domain and range of a few well known functions are defined, let's work a few Examples.

Example 2.5.8: Given the function $f(x) = 5x + 8$, what is the domain and range of this function?

Solution to 2.5.8: This function is an equation of a line, which means that the domain and range are both all real numbers.

Example 2.5.9: Given the function $f(x) = 3 * \sin(x) - 2$, what is the domain and range?

Solution to 2.5.9: In this problem, we have a function of the form $f(x) = a * \sin(x) + d$, with $a = 3$ and $d = -2$. This means that our domain is all real numbers, by the rules stated above. To find the range, plug in our values of a and d into our formula for range. If we do that, the range of this function becomes:

$$[-3 - 2, 3 - 2]$$

Finish computing that to find that the range of this function is [-5, 1]. Therefore, the domain is all real numbers and the range is the interval [-5, 1]

©Brian Gillispie, 2016.

For our last Example, we are going to prove that the range of an exponential function is indeed the interval $(0, \infty)$. This example is advanced, as denoted by the star after the number. If it proves too hard to follow, feel free to skip it and return to it later.

Example 2.5.10*: Prove that the range of $f(x) = 100 * 2^x$ is indeed $(0, \infty)$

Solution to 2.5.10*: To prove this mathematically, we need to show that for any number in the interval $(0, \infty)$, there is some input that will create that output. Therefore, we need to let y be any number in that interval. Since the interval is from zero to infinity, this means that y is positive. Plugging that in gives us the following:

$$y = 100 * 2^x$$

We will now solve this for x, and our solution will be the input that will generate this arbitrary value of y. To solve for x, we need to first divide by 100, which gives us:

$$\frac{y}{100} = 2^x$$

In order to solve this for x, we need to move the x out of the exponent. This is accomplished by taking the logarithm of both sides[38]. Taking the natural log of both sides gives us:

$$\ln\left(\frac{y}{100}\right) = \ln(2^x)$$

This appears to have made the problem harder at first, but there is a rule of logarithms that allows us to move the exponent down. Using the rule of logarithms that states that $\ln(a^x) = x * \ln(a)$, the equation becomes:

---

[38] See Section 6.1 for more on that technique.

©Brian Gillispie, 2016.

$$\ln\left(\frac{y}{100}\right) = x * \ln(2)$$

Divide both sides by $\ln(2)$ to get:

$$\frac{\ln(\frac{y}{100})}{\ln(2)} = x$$

This tells us then, that for any value of $y > 0$, the input $\frac{\ln(\frac{y}{100})}{\ln(2)}$ will generate that output. Therefore, this means that the range of the function $f(x) = 100 * 2^x$ is $(0, \infty)$, as desired.

One thing should be pointed out here. We created a formula to generate a value of x that will yield any value of y. Why is the domain then not $(-\infty, \infty)$ then? The reason has to do with the domain of $\ln(x)$. In the formula, we have the term $\ln(\frac{y}{100})$, which only makes sense provided that $y > 0$. If $y \leq 0$, then the term $\ln(\frac{y}{100})$ is undefined, and there is no input that will generate that output. That then, is why the range is only from $(0, \infty)$ and not from $(-\infty, \infty)$.

©Brian Gillispie, 2016.

2.5 Exercises

For 1 - 5, find the domain of the stated function.

1. $f(x) = -2x + 9$

2. $f(x) = 150 * 0.95^x$

3. $f(x) = \frac{56}{2x+7}$

4. $f(x) = \frac{1}{x}$

5. $f(x) = \sqrt{x - 10}$

For 6 - 21, evaluate the equation at the given point. If the point is outside the domain restriction, report the answer as DNE.

6. $y = 10 * 2^x$, for $x = 10$

7. $y = -16x^2 + 150x$, for $x = 3$

8. $y = 5x + 25$, $0 \leq x \leq 12$, for $x = 5$

9. $y = 50 * 0.7^x$, $0 \leq x \leq 20$, for $x = 3.6$

10. $y = 10x + 250$, $-25 \leq x \leq 25$, for $x = 30$

11. $y = 100 * 1.05^x$, $0 \leq x \leq 5$, for $x = 6$

12. $y = 3.3x - 6$, $2 \leq x$, for $x = 5$

©Brian Gillispie, 2016.

13. $y = -2.3x^2 + 45x + 10$, $0 \leq x \leq 10$, for $x = 2$

14. $y = 100 * 0.97^{x/10}$, $0 \leq x \leq 100$, for $x = 150$

15. $y = 68 + 278e^{0.4t}$, $0 \leq t \leq 5$, for $x = 4.8$

16. $y = 32x + 950$, $0 \leq x \leq 100$, for $x = 86$

17. $y = -1000x + 9000$, $0 \leq x \leq 9$, for $x = 10$

18. $y = 0.35x + 25$, $0 \leq x \leq 50$, for $x = -5$

19. $y = 500 * 1.06^x$, $0 \leq x \leq 30$, for $x = 35$

20. $y = 35 * 0.97^x$, $0 \leq x \leq 60$, for $x = 26$

21. $y = -1.6x^3 + 8x^2 - 3.4x + 10$, $0 \leq x \leq 4$, for $x = 2.27$

22. In the reading we stated that for the function $f(x) = mx + b$, the domain and range is valid provided that $m \neq 0$. Why is $m \neq 0$ important? Look at what happens to the domain and range if m is zero to answer this question.

23* Use the approach in Example 2.5.10 to prove that the domain of the function $f(x) = mx + b$ is the interval $(-\infty, \infty)$, provided $m \neq 0$.

24. For the function $f(x) = 10x + 100$, what is the domain and range?

25. For the function $f(x) = -6x + 60$, what is the domain and range?

26. For the function $f(x) = 1.5^x$, what is the domain and range?

27*. For the function $f(x) = 100 * 1.02^x + 325$, what is the domain and range?

28. For the function $f(x) = -16x^2 + 80x$, what is the domain and range?

29. For the function $f(x) = 33 * \sin(x) + 42$, what is the domain and range?

30. For the function $f(x) = 52 * \sin(x) + 38$, what is the domain and range?

©Brian Gillispie, 2016.

2.6: Application: Ethics and Data Collection

In this Chapter, we have been discussing the various ways a modeler can collect data, as well as the various issues that need to be considered while creating and setting up a model. However, nothing has been said yet about the ethical considerations that need to be kept in mind while collecting data. This Section will give an overview of some of the various issues that a modeler needs to keep in mind while they are collecting data.

2.6.1: Data Collecting Issues

When a modeler wishes to collect data for creating a model, a few issues need to be considered with regards to how that data is collected. First of all, when it comes to humans in a model, a modeler cannot force any human to participate in a study against their will. This means that if someone does not want to participate in your study to find out how eating bananas affects your grades on exams, you cannot force them to participate. This also means that it is not ethical to threaten the person if they refuse to continue the study. In addition, the modeler should make sure that the people participating in the study are aware they can leave at any time, with no consequences.

Secondly, when collecting data, the modeler needs to limit the risk to those they are studying while collecting the data. This means that if you are trying to create a model to see if eating bananas improves test scores, you cannot poison the bananas all of a sudden to see how that works. If your model requires the use of something like that, those participating need to know about the risk before they decide if they wish to participate in the study.

These days, most schools have safeguards in place to make sure that the subjects in a study are subject to the lowest amount of risk required to run the study. For instance, if my purpose in running a study is to see if eating bananas influences your test scores, there is no reason for me to release a wild lion into the testing room. All studies conducted at schools have to pass the approval of a review board to make sure that the mental or physical risk is as low as possible to acquire the results desired.

©Brian Gillispie, 2016.

Now, what if your model requires for something that would put the subjects at high risk? Maybe your model is to determine the effects of a disease outbreak? You are probably not going to be able to acquire volunteers to infect for this study, not to mention the fact that no review board will approve of it either. What modelers often do in this case is they look up previous cases and see what happened, and then use those to create the model. This was used by an engineer in World War II to figure out where to put extra armor on the airplanes [3]. He did not find planes and start shooting at them to collect data for his model. Instead, he collected existing data from planes that had been sent into battle, and then created his model from that data.

Another issue a modeler has to keep in mind when conducting a study is that if the study suddenly becomes dangerous, the modeler has an obligation to abort the study at that point. To see an Example of this, look up the Stanford Prison Experiment [6] of 1971. In this experiment, student volunteers were divided into guards and prisoners, and the researcher wanted to see what would happen to people assigned these roles. Would the person become the role, or would they rise above the role? Turns out that the guards became abusive to the prisoners, and tormented them emotionally to the point that the prisoners forgot they were not in a real prison. The experiment was aborted after 5 days due to how extreme the abuse was becoming, despite the fact that the study was intended to run for 14 days. If, at any point, the data collection is becoming dangerous to those participating, stop the study and see if you can revise it to remove the danger.

A fourth issue that a modeler has to face deals specifically with what are called observational studies. An observational study is a study where the researcher makes no attempt to influence those it is collecting data from, but instead just watches and observes. Most biological models involving finding out the population of a species in an area are observational in nature. However, the fact that the researcher is there does end up impacting the results, through what is known as the Heisenberg Uncertainty Principle. The Heisenberg Uncertainty Principle states that the very act of observing something influences how it acts. For example, if I were to sit in my classroom and just stare at the students as they take a test, I would end up influencing how they act on that test, as I would probably make them nervous and they would not perform as well. Similar arguments could be made that the act of sending a human out to count the wolves in an area might scare off some of the wolves. Or, maybe the human ends up

©Brian Gillispie, 2016.

bringing a virus to the area that he or she is immune to, but it is fatal to the wolves! In these cases, the researcher ended up influencing the results, without intending to.

2.6.2: Data Reporting Issues

Once the data is collected, the modeler has to figure out how to report the data. One of the first rules of science is that the report should be written in such a way that, if they wished, someone could replicate your study. In order for this to happen, the modeler has to report how the data was collected, as well as much of the data collected for the model as it is practical to report. However, the modeler cannot report the data directly as collected, for a couple of reasons.

First of all, the data needs to be reported in a way such that others reading the data do not know the names of who the data was collected from. In other words, the data needs to be kept confidential. Therefore, it is not ethical to report that Laura took 5.2 seconds in the memory trial, while Jake took 10.5 seconds. Instead, the modeler would need to report that subject 1 took 5.2 seconds in the memory trial, and subject 2 took 10.5 seconds. This way, others reading the results still know what the data points are, but do not know who provided those data points.

Secondly, none of the data points can be made up in your model. This should go without saying, but there have been numerous scandals in the past where the data in the study was made up, for the purpose of supposedly supporting a claim. Therefore, to avoid others from suspecting that your data was made up, it is recommended to present the entire data set with your results. This might not be practical for various reasons though, so if that is the case, at least save the data set somewhere and offer to provide it upon request.

However, it is important to remember that in data collection, the modeler is obligated to use all of the data points collected. What if they asked people how many hours they spend watching television each week, and someone reports 10,000 hours? This is an impossible number, but including it in the study would skew any results. What a modeler should do in this case is report that they did indeed get this data point, but since this is an impossible result we will discard it from our study. This way, others reading the paper know that you did get that result, but you are throwing it out. This then makes it possible for someone who disagrees with your assessment of why that point should not be included to go back and rerun your study, with that discarded result included in the data set.

©Brian Gillispie, 2016.

While this list is far from inclusive of the ethical issues involved in collecting and reporting data, it is hoped that this Section at least makes the modeler aware that there are ethical issues to consider. To anyone who intends to create models for scientific and research use, I would recommend consulting with an Institutional Review Board (IRB) to learn more about the ethics and other issues involved with data collection and reporting.

2.6 Exercises

1. One of the more recent ethical issues that have arisen today is the issue of animal subjects in studies. What are some of the issues you can think of that a modeler needs to keep in mind while collecting data on animals?

2. In the reading we mentioned that it is not ethical to force others to participate. However, many schools require that students in general education classes participate in a study, or they lose points off of their final grade. Is this ethical or not, and why?

3. Continuing with Exercise 2, many schools also do allow students to write a paper instead of participating in the study. Is this still ethical, or not, and why?

4. One area that it is difficult to collect accurate data on is questions that are considered controversial in nature. These can be questions that are highly opinionated or questions where a certain answer could land the person in jail. One solution that has been proposed is to have the subjects flip a coin first. If the coin is heads, they tell the truth, if the coin is tails, they lie. The researcher has no knowledge of the result of the coin flip. If this method were used to conduct a study to find the percent of people who drank alcohol when they were under 21, do you think it would be ethical, or not? Why?

5. A researcher collected data from a 14 year old, and then reported the results to their parents. What ethical guidelines listed in this Section did they violate?

6. A teacher posted on their door everyone's exam scores, with the actual names of the students by the scores. What ethical guidelines listed in this section did they violate?

©Brian Gillispie, 2016.

Chapter 2 Summary

2.1: In this Section we discussed the difference between discrete and continuous data, as well as the difference between quantitative and categorical data.

2.2: In this Section, we discussed how to determine the independent and dependent variable. The independent variable is the input of the model, often represented with x. The dependent variable is the output of the model, often represented with y. In addition, we covered how to properly pick your data points from word problems, using these rules.

2.3: In this Section we discussed how one can change the number scale as needed for a mathematical model. Sometimes the change is a change in units, and sometimes it is a change in where in where the model starts on the numerical scale.

2.4: In this Section we discussed how one can assign numbers to categorical data, so that the data can be used in a model. Categorical data can always be assigned numbers, but the numbers need to be assigned in such a way that the number being used has been used by no previous category, and if we add 1 to our assigned number we end up with the number of the next category in order.

2.5: In this Section, we defined the domain and range of a model. The domain is the set of all valid inputs the model can accept, and the range is the set of all valid outputs the model can return. In addition, we discussed how to handle functions that have domain restrictions built into them, as well as what are the domain and range of some commonly known functions.

©Brian Gillispie, 2016.

2.6: In this Section, we discussed some of the ethical issues involved with creating and using data for models. Modelers need to make sure to cause as little harm as possible to their subjects, they need to remember that humans can refuse to participate, and they need to report all data in a way that no one reading it knows the identity of the person or object that provided the data point.

©Brian Gillispie, 2016.

Chapter 3: Measuring Change

To be able to make proper mathematical models, we will need those models to be able to model the change that we are seeing in the data points. In this Chapter we will look at change, as well as various different ways we can calculate change, which will be important for future Chapters of this book. Finally, we will conclude with a couple Sections on how we can use what we know about the change to make short term predictions about the future.

©Brian Gillispie, 2016.

3.1: Measuring Change

Change occurs all of the time in our world. As a result, we are interested in measuring this change, and seeing if we can use it to predict future outcomes. In this Section we will discuss the various ways that we can measure change, so that we can begin to use it to make predictions about the future.

3.1.1: Total Change

We will begin our discussion of change by first considering the total change. The total change is defined as follows:

Definition 3.1.1: The **Total Change (TC)** is defined as the difference between the newest and oldest data point. If we have two points (x_1, y_1) and (x_2, y_2), then the total change can be computed as:

$$TC = y_2 - y_1$$

The units of the Total Change are the same as the units of y.

Let's demonstrate this definition with an Example or two now.

Example 3.1.1: Find the total change between the two data points $(0, 500)$ and $(10, 750)$

Solution to 3.1.1: The total change is defined as the difference between the two y values, which in this case is:

$$TC = 750 - 500 = 250$$

©Brian Gillispie, 2016.

Therefore, the total change is 250.

In the event the data is not in point form, you will have to use the definition of the total change as the difference between the newest and oldest points. Let's demonstrate.

Example 3.1.2: In June 2015, there were 59 ducks on a local pond, and in March 2016 there were 51 ducks on a local pond. Calculate the total change in the number of ducks on the pond.

Solution to 3.1.2: In this case, the data is not in point form, so we need to use the definition of the total change as the difference between the newest and oldest points. Since March 2016 is after June 2015, March 2016 is considered the newest point in this problem, so we are going to calculate the difference between the ducks present in March 2016 and June 2015, which is the following:

$$TC = 51 - 59 = -8$$

Therefore, the total change is -8.

While the total change is quick to calculate, it is not an ideal way to measure change. The reason is, the total change makes no allowance for how time factors in, and would say that the change is the exact same between the points (March 1776, 100), (March 1976, 200) and (March 1975, 100), (March 1976, 200), even though in the first case the change took place over two hundred years, and in the second case the change took place over one year. For that reason, total change is rarely used, and we use other methods to calculate change when modeling.

3.1.2: Average Rate of Change

Another way we can compute change is to calculate the total change, then divide it by how long it takes for the change to occur. This is known as the average rate of change, and can be computed as follows:

> *Definition 3.1.2:* The average rate of change (ARC) is defined as the total change over total time, and can be computed as follows:
>
> $$ARC = \frac{change\ in\ value}{change\ in\ time}$$
>
> The Average Rate of Change has the units $\frac{Units\ of\ Value}{Units\ of\ Time}$

Let's demonstrate this definition now with a few Examples.

Example 3.1.3: A car accelerates 50 mph over 10 seconds. What was the car's average rate of change per second?

Solution to 3.1.3: In this problem, we have that the car's speed is changing by 50 mph, and it takes 10 seconds for the change to occur. Using our formula, that means that the average rate of change is the following:

$$AVC = \frac{50\ mph}{10\ seconds} = 5\ mph/second$$

Or in other words, the car is accelerating at the rate of 5 mph per second.

©Brian Gillispie, 2016.

Example 3.1.4: A researcher has noticed that the number of crickets in the nearby woods has decreased by 1000 over the last 6 months. Compute the average rate of change per month.

Solution to 3.1.4: Be careful here. The question says that the number of crickets is decreasing, so that means there are now -1000 crickets compared to the previous value. So, the change in the value is -1000, and it occurred over 6 months. Plug those into the average rate of change formula to find the following:

$$AVC = \frac{-1000 \text{ crickets}}{6 \text{ months}} = 166.67 \text{ crickets/month}$$

Or, in other words, the number of crickets is decreasing by 166.67 crickets per month, rounded to the nearest hundredth of a cricket.

In these last few Examples, we have seen one way to calculate the average rate of change. However, in all of these cases, we already knew what the change was between the data points, as well as how long it took for the change to occur. In practice this information is provided as two points instead. Therefore, we need to modify our definition to accommodate for the case where two points are given. In the event that two points are provided, the Average Rate of Change is computed as follows:

Definition 3.1.3: The Average Rate of Change (AVC) between points (x_1, y_1) and (x_2, y_2) is defined as follows:

$$AVC = \frac{y_2 - y_1}{x_2 - x_1}$$

Where x represents time and y represents the value you want to find the change of.

We will now use this definition in a few Examples:

©Brian Gillispie, 2016.

Example 3.1.6: A researcher notices that at the start of the month of January there are 25 rabbits, and at the start the month of June there are 55 rabbits. What is the average rate of change in the number of rabbits per month?

Solution to 3.1.6: January is the first month of the year, so we will use the point $(1, 25)$ to mean there are 25 rabbits in January. June is the sixth month, and there are 55 rabbits there so we will use the point $(6, 55)$ to represent that data. Notice in both cases we put the value for time first in our points!

Now with the points found, we will plug in $(1, 25)$ and $(6, 55)$ into our equation. Doing that gives us the following:

$$AVC = \frac{55 - 25}{6 - 1} = 6 \text{ rabbits/month}$$

Notice the units are still units of y divided by units of x in our final answer.

Example 3.1.7: A ball is dropped off of a ledge. When the ball is dropped it is 150 feet above the ground, and 2 seconds later it is 86 feet above the ground. What is the average rate of change of the ball per second?

Solution to 3.1.7: Since the ball is dropped initially, that gives us $(0, 150)$ as our first point. The second point is two seconds after the ball is dropped, which gives us $(2, 86)$ for our second point. Plugging those two points into our formula gives us:

$$AVC = \frac{86 - 150}{2 - 0} = -32 \text{ feet/second}$$

The Average Rate of Change appears ideal for modeling use, as it factors in both the change and the time it took for the change to occur. However, this method is hard to use for problems when the change of the Average Rate of Change is varying as well.

©Brian Gillispie, 2016.

3.1.3: Arithmetic Change

In the event that your data points are sequential, or can be rescaled to be sequential[39], then a special type of change can be computed, which is called the Arithmetic Change. The arithmetic change is defined as follows:

> *Definition 3.1.3:* The **Arithmetic Change (AC)** is defined as the difference between two data points, provided all data points are sequential. In general, if (x_1, y_1) and (x_2, y_2) are two sequential data points, then the Arithmetic Change is defined as follows:
>
> $$AC = y_2 - y_1$$
>
> The units for the Arithmetic Change are the same as the units for the Average Rate of Change.

The calculation for the arithmetic change is the same as the calculation for the total change, with the difference that the x values must be sequential now, or rescaled to be sequential. Let's demonstrate this calculation now in a few examples.

Example 3.1.8: Compute the arithmetic change between the data points $(2, 25)$ and $(3, 30)$.

Solution to 3.1.8: Since our x values are sequential, the arithmetic change is defined. Plugging both of the y values into our arithmetic change formula yields us:

$$AC = 30 - 25 = 5$$

---

[39] To avoid confusion, we will assume that if that was possible, it was already done in this book.

©Brian Gillispie, 2016.

Example 3.1.9: In the month of May, we sold 75 swimsuits, and in the month of June we sold 50 swimsuits. Calculate the arithmetic change on the number of swimsuits sold.

Solution to 3.1.9: June is the month after May, so the data points are sequential. Plugging our number of swimsuits into our arithmetic change formula yields us:

$$AC = 50 - 75 = -25$$

The Arithmetic Change has the advantage that it is easier to calculate than the Average Rate of Change, but has the disadvantage that the points must be sequential for the Arithmetic Change to be able to be computed.

3.1.4: Geometric Change

The last three methods of calculating change are great if you want to know by how much things are increasing over time. But, if you would rather know by what ratio things are increasing by, you need another way to calculate change. After all, most of us would say that an increase of 5 from 5 to 10 means more than an increase of 5 from 10000 to 10005. For this reason, we have one more method that we use to calculate change, and it is the Geometric Change, which is defined as follows:

Definition 3.1.4: The **Geometric Change (GC)** of two equally spaced data points is the ratio of one data point and the data point previous to it. In general, if (x_1, y_1) and (x_2, y_2) are any two sequential data points where $y_1 \neq 0$, the Geometric Change is given by the formula:

$$GC = \frac{y_2}{y_1}$$

The Geometric Change has no units, and is considered unitless for that reason.

©Brian Gillispie, 2016.

Let's demonstrate this definition now with a few Examples.

Example 3.1.10: Compute the geometric change for the data points $(0, 30)$ and $(1, 28)$.

Solution to 3.1.10: Since our points are sequential, the geometric change is defined. Plugging in our y values into the formula yields:

$$GC = \frac{28}{30} = 0.93$$

Therefore, the Geometric Change is 0.93, rounded to the nearest hundredth[40].

Example 3.1.11: In the year 2020 we sold 1500 cars, and in the year 2021 we sold 2000 cars. Compute the Geometric Change of the number of cars sold.

Solution to 3.1.11: Since the year 2021 comes after the year 2020, the data points are sequential and we can compute the Geometric Change. Plugging into the Geometric Change formula the number of cars for each year yields us the following:

$$GC = \frac{2000}{1500} = 1.33$$

This gives us a Geometric Change of 1.33, rounded to the nearest hundredth.

---

[40] While we often wish to avoid rounding due to the errors it causes, when calculating the Geometric Change rounding the final answer is often unavoidable.

©Brian Gillispie, 2016.

Example 3.1.12: At 7 am the outside temperature is 0 degrees, and at 8 am the outside temperature is 4 degrees. Compute the Geometric Change for the outside temperature.

Solution to 3.1.12: The times are sequential, as 8 am is the hour after 7 am, so we can compute the Geometric Change. Plugging into our Geometric Change formula gives us the following:

$$GC = \frac{4}{0}$$

However, division by 0 is undefined, so the Geometric Change[41] is undefined in this case.

---

[41] Some say the Geometric Change is ∞ when this happens instead.

©Brian Gillispie, 2016.

3.1 Exercises

For 1 - 9, compute the Total Change (TC) and the Average Rate of Change (AVC). State the units of your answer if known.

1. $(0, 100)$ and $(5, 70)$

2. $(10, 500)$ and $(20, 1000)$

3. A colony of frogs had 20 frogs in February 2015, and 70 frogs in July 2015.

4. At 2 am there was 1 inch of snow on the ground, at 8 am there was 5.2 inches of snow on the ground.

5. On May 12th there were 1000 crickets, and on May 16th there were 1960 crickets.

6. For the home opening football game, the school sells 45,672 tickets. For the sixth home game of the season, the school sells 82,521 tickets.

7. In 1985, the newest game systems retailed for $199. Now in 2013, the newest game systems are retailing for $499.

8. The cost of gas was $0.99 at the beginning of 1994, and it is now $3.09 at the beginning of 2014.

9. On June 16th you could buy milk for $3.99 a gallon, and on June 23rd milk was $3.79 a gallon.

For 10 - 21, compute the arithmetic change (AC) and the geometric change (GC)

10. $(10, 50)$ and $(11, 65)$

11. $(2006, 7.2)$ and $(2007, 8.8)$

12. $(12, 55)$ and $(13, 100)$

13. (October 2086, 1500) and (November 2086, 1550)

14. A store notices that it sold 12 winter coats in February of 2017 and 5 winter coats in March of 2017.

15. A pond has 54 ducks in March 2015 and 66 ducks in April 2015.

16. Your favorite sports team won 2 games in the year 2050, and in 2051 they won 7 games.

17. On May 16th you sold 5 credit cards, and on May 17th you sold 8 credit cards.

18. On Friday the store had 752 customers, on Saturday the store had 818 customers.

19. Your school football team scored 0 points during the first game of the season, and 17 points during the second game of the season.

20. Your school baseball team scored 4 runs during the tenth game of the season, and 4 runs during the eleventh game of the season.

21. An island has 150 goats in the year 2025, and 135 goats in the year 2026.

©Brian Gillispie, 2016.

3.2: Interpretation of Change

Now that we have discussed the various ways we can calculate change between two data points, we will now discuss how to interpret our answers? This is important because knowing the valid interpretation of our answer will make it possible to better understand what is going on in the data set that is being modeled.

3.2.1: Interpreting Average Rate of Change

In order to better understand how Average Rate of Change works, let's open up with an Example. Pretend you have been monitoring how many customers have been coming to a waterpark during the early part of the summer. You record that on May 27th, there were 1450 customers that day, and on May 28th, there were 1600 customers that day. Plugging those numbers into the Average Rate of Change formula (and using 27 for May 27th and 28 for May 28th), we get the following:

$$AVC = \frac{1600 - 1450}{28 - 27} = 150 \text{ customers/day}$$

Or, our Average Rate of Change is 150 customers per day, based on these numbers. Now, with that number found, let's try something else. If we take our number of customers on May 27th, which was 1450, and add 150 to that number, notice what we get:

$$1450 + 150 = 1600$$

Which, incidentally, is the number of customers we had on May 28th, or the next day in the sequence! It turns out that you can always do this; you can always take one data point and add the Average Rate of Change to it, and get the next number in the sequence[42]! For this reason, the Average Rate of Change is often interpreted to mean how much change occurs, as we

---

[42] Be careful, that may not be the same as the next data point. If we had used May 27th and May 30th, we'd still only get May 28th's number of customers this way.

©Brian Gillispie, 2016.

increase time by one unit. In other words, this means that if we had an Average Rate of Change of 5 goats/day, that would mean that as each day passes the number of goats in the area increases by 5. For this reason, the Average Rate of Change is often interpreted as the amount of change that occurs, per unit of time, and therefore, any time you need to know the amount of change that occurs per unit of time, you always calculate the Average Rate of Change.

Incidentally, this interpretation is also valid for the Arithmetic Change, as the Arithmetic Change is just the Average Rate of Change when the time units are sequential. For that reason, we will often use the Arithmetic Change if the time units are sequential and the Average Rate of Change if the time units are not sequential. Either way, the interpretation is the same in both cases.

3.2.2: Interpreting Geometric Change

Like what we did with Average Rate of Change, we will open up with an Example. Assume that you are in charge of reporting how many bees are in a beehive, and in the month of May you notice 10000 bees in the beehive, and in the month of June you notice 11200 bees in the beehive. Plugging those numbers into the Geometric Change formula gives us the following:

$$GC = \frac{11200}{10000} = 1.12$$

Or, our Geometric Change is 1.12, based on those numbers.

Now, consider another question. This time, you wish to find the percent the number of bees in June is of the number of bees in May. The formula to find the percent one number is of the second number is to divide the first number by the second, then convert that number into a percent. In other words, to find the percent the number of bees in June is of the number of bees in May, we have to divide 11200 by 10000, then covert our number to a percent. If we do that, we end up with the following:

$$\frac{11200}{10000} = 1.12$$

©Brian Gillispie, 2016.

Then, we convert our answer of 1.12 to a percent to get 112%. So, our bees in June are 112% of the number of bees we had in May.

The reader may notice that while calculating the percent the number of bees in June is of the number of bees in May, we ended up with 1.12 at one point, which is the same as our Geometric Change. It turns out that this is not a coincidence, as the formula for calculating the Geometric Change is very similar to the formula for computing the percent two numbers are of each other. In fact, the only difference is that the Geometric Change gives the answer in decimal form, and the percent of formula gives the answer in percent form. This means that we can then convert our Geometric Change to a percent, and use it to tell us how much our second data point is of the first.

However, most people prefer to know the percent increase or percent decrease instead of the percent of. Therefore, to convert our Geometric Change to a percent increase or percent decrease, we need to figure out by how much our original amount has changed. The easiest way to find the percent increase of two numbers is to subtract 100% from the percent form, or subtract one if it is in decimal form. Then, if the new answer is positive, it is a percent increase, and if it is negative, it is a percent decrease.

3.2.3: Putting it All Together

Now that we have seen how the different changes[43] can be interpreted, let's work a few Examples where we explore this more. Remember that if you need to know the amount of change per unit of time, compute the Average Rate of Change. If you need either the percent two numbers are of each other, or the percent increase or decrease, compute the Geometric Change. Let's demonstrate this now.

---

[43] Except Total Change, which is hard to interpret and use well.

Example 3.2.1: In the month of March, there were 25 turtles on an island, and in the month of June there were 46 turtles. Calculate the Amount of Change per unit of time in this situation.

Solution to 3.2.1: Since we want the Amount of Change per unit of time, and the time units are not sequential, we will calculate the Average Rate of Change, using 3 for March and 6 for June. Doing that gives us the data points of (3, 25) and (6, 46), which, if we plug those into the formula for Average Rate of Change, we end up with:

$$\frac{46 - 25}{6 - 3} = 7 \text{ turtles/month}$$

This means that the Amount of Change per unit of time is 7 turtles per month.

Example 3.2.2: In the month of January, a car dealership sold 87 cars. In the month of February, a car dealership sold 95 cars. By what percent did the sales of cars increase in this time?

Solution to 3.2.2: This time we want to find a percent increase, so we need the Geometric Change. Using our Geometric Change formula, we end up with:

$$\frac{95}{87} = 1.0920$$

Now, to find the percent increase, we need to subtract one from this number. If we do that, we end up with:

$$1.0920 - 1 = 0.0920$$

Now, convert that to a percent to finish. This means that in February the number of sales increased by 9.20%

©Brian Gillispie, 2016.

Example 3.2.3: A copier place currently spends $5600 on maintenance per month. However, after some cost saving measures are implemented, they now spend $5250 on maintenance per month. By what percent did the cost of maintenance decrease due to these cost saving measures?

Solution to 3.2.3: Since we wish to find a percent decrease, we need to calculate the Geometric Change. Using the formula for Geometric Change with our data points, we end up with:

$$\frac{5250}{5600} = 0.9375$$

Now, we need to subtract one from that number, which if we do that, we end up with:

$$0.9375 - 1 = -0.0625$$

Convert to a percent and drop the sign[44] to end up with our percent decrease is 6.25%.

Example 3.2.4: An observatory has reported sighting 5 whales in the month of October, and 4 whales in the month of November. What is the Amount of Change per unit time for the number of whale sightings?

Solution to 3.2.4: Since our data points are sequential months, we can use the Arithmetic Change formula instead of the Average Rate of Change formula. Doing that gives us:

$$4 - 5 = -1$$

Therefore, the Amount of Change per unit of time is -1 whales per month[45].

---

[44] The reader may have seen percents reported with negative signs in front of them. When the sign is still reported, the percent is being reported as percent change, not as a percent of decrease.

[45] Recall the units of the Arithmetic Change are still in units change of value/change in time.

©Brian Gillispie, 2016.

Example 3.2.5: Using the same data in Example 3.2.4, by what percent is the number of whales sighted decreasing?

Solution to 3.2.5: Since we want a percent, we need to compute the Geometric Change. Plugging these numbers into our formula for Geometric Change, and recalling that 4 whales were sighted in the later month, gives us the following:

$$\frac{4}{5} = 0.80$$

Then, subtract one and convert the answer to a percent (remember to drop the sign!), and you would end up with that the percent of decrease in whale sightings is 20%, based on this data.

©Brian Gillispie, 2016.

3.2 Exercises

1. In the month of May a popular waterpark saw 10,800 customers, and in the month of August the same waterpark saw 22,500 customers. Calculate the amount of change over time.

2. In 2015 a wildlife reserve reports 250 ducks living on the reserve, and in 2016, the same reserve reports 300 ducks living on the reserve. Calculate the amount of change over time.

3. In the year 2002 it cost $2700 for tuition for a semester at a school, and in 2010 it cost $5400 for tuition at the same school. Calculate the amount of change over time.

4. In the month of October 2015 a wildlife reserve reported 200 ducks on the reserve, and in the month of January 2016, the same reserve reported 5 ducks on the reserve. Calculate the amount of change over time.

5. A training crew reported 122 sales in previous week, and 155 sales in the current week. Calculate the percent increase in sales.

6. In May your favorite candy bar cost $1.59, and in the month of June the same candy bar cost $1.69. Calculate the percent increase in the cost of the candy bar.

7. Last week a gallon of gas cost $2.99, and this week it now costs $2.79. Calculate the percent decrease in the cost of the gallon of gas.

8. In July you owned $5000 worth of stock, and in August you owned $5060 worth of stock. Calculate the percent increase in the amount of stock owned.

9. Last year you made $10.25 an hour, and this year you make $10.88 an hour. Calculate the percent increase in your hourly salary.

©Brian Gillispie, 2016.

10. In October a waterpark reported 764 customers, and in November, the same waterpark reported 112 customers. Calculate the percent decrease in the number of customers.

11. A wildlife reserve reports 368 geese on the reserve in the year 2019, and 362 geese on the reserve in the year 2020. Calculate the percent decrease in the number of geese.

12. On January 29th, you had 8 sales, and on January 30th you had 6 sales. Calculate the percent decrease in the number of sales.

13. On March 6th there is 10 liters of pollutant reported in a lake. On March 7th there is 9.4 liters of pollutant reported in the lake. Calculate the percent decrease in the amount of pollutant in the lake.

14. In the year 2016 there are 50 wolves on an island, and in the year 2017 there are 60 wolves on the same island. Calculate the percent increase in the number of wolves on the island.

15. Last month you had a horrible month on the sales floor, and only sold 15 items. Your boss offers you a bonus of $200 provided you increase your sales by 100% next month. Next month you sell 32 items. Do you get the bonus? Why or why not?

16. A manufacturing plant was ordered to reduce the cost of production by 10% over the course of the year. Last year the cost of production was $125,000, and this year the cost of production was $115,000. Did they meet the goal? Why or why not?

17. A college wishes to increase its enrollment by 5% for the next year. In the year 2017 they had 2560 students, and in 2018 they had 2600 students. Did they meet the goal? Why or why not?

©Brian Gillispie, 2016.

3.3: Using Change: Single Point Modeling

Now that we know how to calculate and interpret change, we now direct our attention to how to use change to make predictions about the future. In this Section, we are going to focus on problems where we know one data point, and the change for the next data point is already known. Because we only know a single data point, this type of modeling is often called Single Point Modeling for that reason.

3.3.1: Using the Average Rate of Change

In the last Section, we saw that the Average Rate of Change tells us the amount of change per unit of time. What this means is if we know the Average Rate of Change, and the current value, we can add the Average Rate of Change to the current value and find the next value. However, you do have to keep in mind that when you do this you will only find the data point for the next point in time, so you may have to add the Average Rate of Change multiple times to find the desired value. Let's demonstrate all of this with a few Examples now.

Example 3.3.1: A wildlife reserve has 50 goats, and the Average Rate of Change for the goats is currently 10 goats per year. Predict how many goats will be present next year.

Solution to 3.3.1: We currently have 50 goats, and we know that the number of goats is changing by 10 goats per year, therefore the number of goats present next year will be:

$$50 + 10 = 60$$

Example 3.3.2: At 2 pm the local lake is 20 feet high, and the Average Rate of Change of the height of the lake is 0.1 feet per hour. How high will the lake be at 4 pm?

Solution to 3.3.2: This time you have to be careful, because adding the Average Rate of Change to the height of the lake at 2 pm will only tell you the height of the lake at 3 pm. For that reason,

©Brian Gillispie, 2016.

you have to add twice, once to get the 3 pm height, and once to get the 4 pm height. If you do that, you will end up with the following for the lake height at 4 pm:

$$20 + 0.1 + 0.1 = 20.2 \text{ feet}$$

Notice how we had to add twice in the second Example. The reason is the Average Rate of Change was in units of feet per hour, which means that each time you add the Average Rate of Change; you will increase the time in the model by one hour! Always check what units your Average Rate of Change is in, because it will tell you what time steps you will be advancing in when working the problem, as this next Example will demonstrate.

Example 3.3.3: At 5 pm there was 6.8 inches of snow on the ground, and the Average Rate of Change for the snowfall is 1.2 inches/2 hours. How much snow will be on the ground at 9 pm?

Solution to 3.3.3: Careful! The units on the Average Rate of Change is in inches per 2 hours, which means every time you add the Average Rate of Change, you will increase time by 2 hours. Therefore, the first time you add 1.2 you will jump from 5 pm to 7 pm, as 7 pm is two hours after 5 pm. Therefore, to solve this problem, we need to add twice[46], once to go from 5 pm to 7 pm, and once more to go from 7 pm to 9 pm. This means our final answer is:

$$6.8 + 1.2 + 1.2 = 9.2 \text{ inches}$$

Usually when information is given in problems it is not in the words Average Rate of Change, but in other language instead. When that happens, look at problem closely to see if you know an amount of change over time. If you do, treat that as an Average Rate of Change. Let's demonstrate with an Example.

---

[46] Alternatively you could convert the Average Rate of Change to inches per hour, then add that result 4 times instead.

©Brian Gillispie, 2016.

Example 3.3.4: A sales floor had 250 sales for the month of June, and management wants the number of sales to increase by 1 sale per month. How many sales should the floor try to have for the month of July?

Solution to 3.3.4: This time the words Average Rate of Change are not present, but we are given a change amount per unit of time, which is the same thing. Notice we are told in the problem an increase of 1 sale per month is desired. This translates to an Average Rate of Change of 1 sale per month. Then, since we need to predict the sales goal for next month, you add the Average Rate of Change to the sales from June to get July's sales goal, which is:

$$250 + 1 = 251$$

Example 3.3.5: On Monday your work commute was 45 minutes, and you wish to reduce it by 2 minutes per day for as long as it is possible to do so. How long do you want your work commute to be on Tuesday?

Solution to 3.3.5: This time we are given an amount of change, which is a reduction of 2 minutes per day. This translates to an Average Rate of Change of -2 minutes per day. Then, to get Tuesday's goal, we need to take Monday's time and add our Average Rate of Change to it, which gives us that Tuesday's goal is:

$$45 + (-2) = 43 \; minutes$$

Notice we used a negative Average Rate of Change in the last example. This is due to the fact that we wanted to reduce our commute time, which requires the amount of change to be negative. Always be on the lookout for words which imply that the change might be negative, as those will help you to know which sign to use on the Average Rate of Change in these types of problems.

©Brian Gillispie, 2016.

3.3.2: Using the Geometric Change

Now that we have seen how to use the Average Rate of Change to make a prediction, let's look at the Geometric Change next. Recall that the Geometric Change by definition is the following:

$$GC = \frac{y_2}{y_1}$$

Multiply both sides by y_1 and you have instead:

$$y_1 * GC = y_2$$

What this means is if you know the Geometric Change (GC) and you know a current value, you can predict the next value by multiplying the previous value by the Geometric Change. Let's demonstrate.

Example 3.3.6: In 2017 there are 180 rabbits on an island, and the Geometric Change is 2.5. Predict how many rabbits there will be present in 2018.

Solution to 3.3.6: Since we know the current number of rabbits, and the Geometric Change, we multiply the two numbers to get our prediction for the number of rabbits in 2018, which is:

$$180 * 2.5 = 450 \text{ rabbits.}$$

Be careful when doing this though, as the Geometric Change will only find the value for the next unit of time[47], as this next Example will show.

---

[47] Since the Geometric Change is unitless, it is always assumed that you are working in whatever units are relevant to the problem.

©Brian Gillispie, 2016.

Example 3.3.7: After taking a medicine at 1 pm, there is 325 mg of the medicine in the body, and the Geometric Change is 0.93. How much medicine is in the body at 3 pm?

Solution to 3.3.7: Since we need to advance two time steps now, we need to multiply by the Geometric Change twice to find the amount of the medicine in the body at 3 pm. Doing that yields the following:

$$325 * 0.93 * 0.93 = 281.0925 \text{ mg}$$

Usually when the change is known, it is not presented using the words Geometric Change. Instead the change is either presented in the form percent of, percent increase, or percent decrease. When those happen, you will need to convert the answer to the Geometric Change first, and then multiply. Let's demonstrate with a few Examples.

Example 3.3.8: You are currently making $50000 a year in the year 2019, and will get a 3% raise in 2020. How much is your new salary?

Solution to 3.3.9: This time, you are given the percent increase, and need to covert that to the Geometric Change before proceeding. To convert a percent increase to the Geometric Change, first convert it to a decimal, and then add one to it. When you do that, you will get that your Geometric Change is:

$$GC = 0.03 + 1 = 1.03$$

Next, multiply your given amount by the Geometric Change to finish. When you do that, you will find that your salary in 2020 will be:

$$50000 * 1.03 = 51500$$

©Brian Gillispie, 2016.

Example 3.3.10: You wish to buy a computer, which costs $1299. Currently the store is having a sale and all computers are 20% off[48]. How much will a computer cost you today due to the sale?

Solution to 3.3.10: Again, we need to convert the percent to the Geometric Change before proceeding. This time we have that the percent is a 20% decrease (as percent off is a percent decrease by definition), so first covert the percent to a decimal. Then, since it was a percent decrease, we need to add on a negative sign. Then, add one to it to get the Geometric Change. If we do that, we get that our Geometric Change is:

$$GC = -0.20 + 1 = 0.80$$

Then, multiply the cost of the computer by the Geometric Change to get the final cost of the computer, which is:

$$\$1299 * 0.80 = \$1039.20$$

Example 3.3.11: In the year 2018 you are working for a company which pays you $13 an hour. Each year your salary will increase by 4% while you are at this company. How much will you be making an hour in 2021?

Solution to 3.3.11: Careful! A common error is to treat the total increase as 12% here and go from there. However, you cannot do that as you **cannot** add percents for these kinds of problems. Instead, you have to treat the increase as 4% for each year, find the Geometric Change, and then multiply your current salary by the Geometric Change three times to find how much you will make in 2021.

So, first, let's find the Geometric Change. Since our salary is increasing, that means we first convert the percent to a decimal, and then add one. If we do that, we end up with the following for our Geometric Change:

---

[48] Not to be confused with 20% of, which means something completely different.

©Brian Gillispie, 2016.

$$GC = 0.04 + 1 = 1.04$$

Then, we multiply our salary by 1.04 three times, as we need to calculate the value three years into the future. Doing that gives us our final answer of:

$$\$13 * 1.04 * 1.04 * 1.04 = \$14.62 \text{ an hour}$$

3.3.3: Combined Change

Sometimes both an Average Rate of Change and a Geometric Change is applied to the same data point. Unless we are told otherwise, the way we handle these situations is to apply the Geometric Change first, then the Average Rate of Change is applied to the answer[49]. Let's demonstrate this with a couple of Examples to close out this Section.

Example 3.3.12: You are currently making $11.65 an hour, but have just been awarded a raise of $1 an hour, plus an 3% increase for cost of living. What is your new hourly salary?

Solution to 3.3.12: A 3% increase means that the Geometric Change is 1.03, and a raise of $1 an hour means that the Average Rate of Change is $1. Since the Geometric Change is applied first, we take our hourly salary and multiply it by the Geometric Change, and then add the Average Rate of Change to our answer[50]. This results in the following:

$$11.65 * 1.03 + 1 = \$13 \text{ } an \text{ } hour$$

Therefore, your new salary is $13 an hour.

---

[49] The exception is coupons. Usually if you have a $5 off coupon and a 20% off coupon, they will process the $5 off first (the Average Rate of Change) then the 20% off coupon (the Geometric Change).

[50] Technically, when you do this, you are saying the Geometric Change from point 1 to point 2 is 1.03 then the Average Rate of Change from point 2 to point 3 is $1. Point 3 is your final answer.

©Brian Gillispie, 2016.

Example 3.3.13: A river has 15000 salmon in it in the year 2018. Each year the number of salmon in the river increase by 5%, and harvesting cuts the number of salmon in the river by 750 per year. How many salmon will be in the river in the year 2019?

Solution to 3.3.13: To solve this we need to first apply the Geometric Change of 1.05, due to the 5% increase, and then apply the Average Rate of Change of -750, due to harvesting. Applying both of these in order means that the number of salmon in the river in the year 2019 is:

$$15000 * 1.05 - 750 = 15000 \text{ salmon}$$

Therefore, in the year 2019 there will still be 15000 salmon, the same as there were in 2018.

3.3 Exercises

1. A store had 2000 customers on May 21st, and they know that the Average Rate of Change is 20 customers/day. Predict how many customers they will have on May 22nd.

2. You ran a mile in 600 seconds on August 3rd, and due to training the Average Rate of Change in your running time is -1 second per day. Predict how fast you will run the mile on August 4th.

3. In the year 2020 there are 300 rabbis on an island, and the Geometric Change in the number of rabbits is 1.75. Predict how many rabbits will be present in the year 2021.

4. A radioactive isotope has 250 mg present on October 3rd, and the Geometric Change in the number of mg present is 0.65. Predict how many mg will be present on October 4th.

5. You are making $12 an hour in the year 2024, but your job promises to give you a raise of $1 per year. How much will you be making in the year 2025?

6. Repeat Exercise 5, but now predict how much you will be making in the year 2027.

7. A pond has 10 fish present on March 16th, and every day you are adding 5 fish to the pond. How many fish will be present on March 19th?

8. On March 16th, the same pond has 100 frogs, and the number of frogs present is increasing by 4% per day. How many frogs will be present on March 19th?

9. A lake has 100 ml of a pollutant present on July 4th, and filtering processes are reducing the amount of pollutant present by 10% per day. How much will be present on July 6th?

10. While training, you notice that on September 19th you ran a mile in 550 seconds, and your time is dropping by 12 seconds per week. How fast will you run a mile on September 26th?

©Brian Gillispie, 2016.

11. You wish to buy a shirt that is $25, and everything in the store is 35% off. How much will the shirt cost you?

12. An island has 30 deer on it in the year 2019, and the number of deer is increasing by 10% per year. How many deer will be present in the year 2021?

13. You want to buy a lunch that costs you $7, and you have two 20% off coupons. How much will your lunch cost you?

14. Repeat Exercise 13, but this time assume you have five 20% off coupons. How much will your lunch cost you? *Hint: The answer is not $0.*

15. You decide to buy a computer that normally costs $1300, but it is on sale for 80% off. In addition, you have $300 in store credit you can use to buy the computer after the sale is applied. How much will the computer cost you?

16. A fishery has 1300 catfish in the year 2011, and the number of catfish grows by 10% per year. In addition, 100 catfish are harvested every year. How many catfish will be present in the year 2012?

17. You are currently making $10 an hour in the year 2014, and your job offers you a 2% cost of living increase, and an $1 an hour raise. How much will you be making after both raises take effect?

18. A restaurant has 1600 customers on January 21st, and projections say that the number of customers will increase by 10% tomorrow. However, a big snowstorm occurs on January 22nd which results in 980 customers staying home due to the weather. How many customers arrive on January 22nd?

©Brian Gillispie, 2016.

19. An island has 10000 rabbits in the year 2015, and the number of rabbits is increasing by 250% per year. In an effort to reduce the rabbit population, rabbit hunting is allowed on the island, which reduces the population by 20000 rabbits. How many rabbits will be present in the year 2016?

20. Repeat Exercise 19, but now predict how many rabbits will be present in the year 2017.

21. A pizza place sold 150 pizzas on July 3rd, and projections indicate that the number of pizzas they will sell will decrease by 65% for the next day. However, on July 4th, a huge party orders an extra 100 pizzas not covered in the projections. How many pizzas will they sell on July 4th now?

3.4: Using Change: Two Point Modeling

Now that we have worked with the Average Rate of Change and the Geometric Change when only a single data point is known, we will now discuss how to handle the situation where two data points are known. Like we did when only one data point is known, we will discuss how to handle the Average Rate of Change and the Geometric Change in turn.

3.4.1: Using the Average Rate of Change

To begin our discussion of Average Rate of Change and two points, let's consider an example. Pretend that you have been asked to create a mathematical model that will be used to predict the number of wolves in a forest, after 30 wolves were released in the year 2009. In 2010, you count the number of wolves again, and discover that there are now 33 wolves in the forest. Based on these observations, you would calculate the Average Rate of Change to be 3 wolves per year.

Now, pretend we wish to know how many wolves will be present in the year 2011. A reasonable assumption is that our Average Rate of Change will not change significantly from 2010 to 2011, or in other words, we should expect that the Average Rate of Change will still be about 3 wolves per year. Therefore, we can estimate the number of wolves present in 2011 by taking the number of wolves present in 2010, and adding the Average Rate of Change to them. If we do that, we end up with a prediction of 36 wolves present in the year 2011.

Notice what we did there. In our example, we took the Average Rate of Change of the previous two data points and added it to the y value in the second point, to predict the y value for a third data point! This is how the Average rate of Change is often used to make predictions, by calculating the Average Rate of Change for known points, then predicting from there. Let's demonstrate this method with a few more Examples now.

Example 3.4.1: In May there were 12 ducks on the pond, and in June there were 15 ducks. Using the Average Rate of Change, predict how many ducks will be present in July.

Solution to 3.4.1: To begin, we need to know the Average Rate of Change between our two points. We know that there were 12 ducks on the pond in May, and 15 ducks in June. Therefore, our two points should be $(5, 12)$ and $(6, 15)$.

With our two points now found, we can calculate the Average Rate of Change. The Average Rate of Change (ARC) for these two points is:

$$ARC = \frac{15 - 12}{6 - 5} = 3 \text{ ducks/month}$$

Finally, add that Average Rate of Change to the y value of the last point to predict the next y value. Since we know there are 15 ducks in June, there should then be $15 + 3$ or 18 ducks in July. Therefore, we would say that we estimate there will be 18 ducks in July, based on our numbers.

Example 3.4.2: A company wishes to predict sales of the newest video game for this week. If they sold 1250 copies of the game on Monday, and 1500 copies on Wednesday, using the Average Rate of Change, how many copies should they expect to sell on Thursday?

Solution to 3.4.2: Start off by identifying what our two points are for this problem. We know there were 1250 copies sold on Monday and 1500 copies on Wednesday, so if we let 1 represent Monday and 3 represent Wednesday, our points are then $(1, 1250)$ and $(3, 1500)$.

Next, compute the Average Rate of Change. Plugging these points into the Average Rate of Change formula yields the following for our Average Rate of Change:

$$ARC = \frac{1500 - 1250}{3 - 1} = 125 \text{ copies/day}$$

To finish, we need to add our Average Rate of Change to the number of copies sold on Wednesday, which means that we should expect to sell on Thursday:

©Brian Gillispie, 2016.

$$1500 + 125 = 1625 \text{ copies}$$

Remember that when you add the Average Rate of Change, you only advance forward one time unit. Sometimes this means you will have to add the Average Rate of Change multiple times in order to reach the desired point, as this next Example will demonstrate.

Example 3.4.3: A state wants to create a math model to predict population growth in an area. They know that a town has 250 students in the year 2000, and 275 students in the year 2001. Using the Average Rate of Change, predict how many people will be present in the year 2003.

Solution to 3.4.3: Again, we begin by finding our two points. We know that the population was 250 in the year 2000, and 275 in the year 2001, so our points will be $(2000, 250)$ and $(2001, 275)$ for this problem. Now, calculating the Average Rate of Change for those points gives us:

$$ARC = \frac{275 - 250}{2001 - 2000} = 25 \text{ people/year}$$

However, notice that the Average Rate of Change is in the units people per year, which means every time you add it, you will only jump forward one year. This means to go from 2001 to 2003 you will need to add the Average Rate of Change two times to the number of people in 2001. Doing that gives us that in 2003 the town should have:

$$275 + 25 + 25 = 325 \text{ people}$$

Because of the way the Average Rate of Change works, you have to pay careful attention to your units, as sometimes the time steps when you add the Average Rate of Change isn't the same as it was for the data points. Look again at Example 3.4.2. In that Example, the data points were two days apart, but the Average Rate of Change was still in customers per day. Because of that, when we added it, we only went forward one day, not two. You always have to pay attention to the units, else you may add too many or too few times, and end up with a bad prediction.

Let's now close this discussion on how to use the Average Rate of Change with another Example, where this time we will have to pay close attention to the units of our Average Rate of Change.

Example 3.4.4: A town notices that it has 1650 people in the year 2000 and 1500 people in the year 2010. How many people will the town have in the year 2013?

Solution to 3.4.4: Start off by finding your two points. Since we know that there are 1650 people in the year 2000 and 1500 people in the year 2010, that makes our points for this problem (2000, 1650) and (2010, 1500).

Next, plug those points into the Average Rate of Change formula. Doing so yields the following for our Average Rate of Change:

$$ARC = \frac{1500 - 1650}{2010 - 2000} = -15 \text{ people/year}$$

Since the Average Rate of Change is in people per year, and we need to predict how many people are present three years later, we need to add the Average Rate of Change three times to our known population in the year 2010. If we do that, we end up with the following for our prediction for the number of people present in the year 2013:

$$1500 - 15 - 15 - 15 = 1455 \text{ people}$$

3.4.2: Using the Geometric Change

Let's open up our discussion of using Geometric Change and two data points by returning to our wolf example. Recall that we had 30 wolves in 2009 and 33 wolves in 2010. Based on those points, you would calculate the Geometric Change to be 1.1 for those data points.

Now, this time, pretend that you wish to use the Geometric Change to predict how many wolves will be present in 2011. Since the Geometric Change was 1.1 from 2009 to 2010, it seems reasonable to assume that the Geometric Change will be 1.1 also from 2010 to 2011. Therefore, we can take our number of wolves in 2010 and multiply them by the Geometric Change, which will give us that there will be 36.3 wolves present in 2011, or 36 wolves after we round to the nearest whole number.

Notice that the idea is the same as it was with Average Rate of Change. We take our Geometric Change for our two data points and multiply the y value of our second data point by the Geometric Change to find our next predicted value. Let's demonstrate this now with a few Examples.

Example 3.4.5: A small town is slowly losing its population as people move away to the big city. In the year 1990 the town had 675 people, and in the year 1991 the town had 663 people. Use the Geometric Change to predict how many people will be there in the year 1992.

Solution to 3.4.5: First, we need to figure out what our points are. In this problem, we have the points (1990, 675), and (1991, 663). Using our definition of Geometric Change, we calculate the Geometric Change (GC) as follows:

$$GC = \frac{663}{675} = 0.98\overline{2}$$

To finish, we need to multiply our Geometric Change to the number of people in the year 1991. If we do that, we end up with the following for the number of people present in 1992:

$$663 * 0.98\overline{2} \approx 651 \text{ people}$$

©Brian Gillispie, 2016.

If you need to go further out than one data point into the future, remember that we can do that by multiplying by our Geometric Change multiple times. Let's demonstrate.

Example 3.4.6: In 1859 Thomas Austin[51] released 24 rabbits onto his farm in Western Australia, wanting something to hunt. In 1860 it is estimated that there were now 150 rabbits present in Australia. Predict how many rabbits will be present in 1862 in Australia, using the Geometric Change.

Solution to 3.4.6: Again, we start out by identifying what the two data points are. In this problem, our data points are $(1859, 24)$ and $(1860, 150)$. Then, using our definition of Geometric Change, we calculate the Geometric Change as follows:

$$GC = \frac{150}{24} = 6.25$$

Since we wish to know how many rabbits are present in 1862, we need to multiply the number of rabbits in 1860 by the Geometric Change twice. Therefore, this means that in the year 1862, we estimate there were:

$$150 * 6.25 * 6.25 \approx 5{,}859 \text{ rabbits}$$

Since the Geometric Change requires that the data points either be sequential or able to be scaled to be sequential, we will not cover how to handle problems where the data points are not sequential, as those are usually solved via the Average Rate of Change.

---

[51] This really happened [11], and as it turns out that Thomas forgot that rabbits had no natural predators in Australia, so they grew really fast. How fast you may ask? Between 1865 and 1866 Thomas alone killed more than 34,000 rabbits on his property alone! In 1990 it was estimated that there were over 600 million rabbits in Australia.

©Brian Gillispie, 2016.

3.4.3: Putting it all Together

In all of our Examples so far, we were told which method to use to make the desired prediction. In reality, that is not normally the case, and the modeler has to figure out which method should be used to make the prediction. In this case, the modeler will often[52] run both models, and then report the range between the results[53]. Let's demonstrate this technique now.

Example 3.4.7: A researcher is interested in learning how many squirrels there are in a city park. In the month of May, the researcher estimates there were 250 squirrels, and in the month of June the researcher estimates there were 275 squirrels. Estimate how many squirrels there will be in the month of July.

Solution to 3.4.7: This time, we are not told which method to use, just to make an estimation somehow. Nevertheless, the technique is still the same. Start out by figuring out what are the given data points. In this problem, we are given the two points $(5, 250)$ and $(6, 275)$. Plugging those into our Average Rate of Change (ARC) and Geometric Change (GC) formulas yields the following:

$$AC = 275 - 250 = 25$$

$$GC = \frac{275}{250} = 1.1$$

Therefore, with these two numbers found, we can make a prediction of how many squirrels there will be present in July. Using the Average Rate of Change, we find that there will be $275 + 25$ squirrels, or 300 squirrels. Using the Geometric Change, we find that there will be $275 * 1.1$ squirrels, or 303 squirrels, rounded to the nearest squirrel.

Therefore, since the two numbers are not that far apart, it would be reasonable to predict that there should be between 300 and 303 squirrels in the month of July.

---

[52] Unless there is a way to eliminate one of the models that is. However, this is very rare in reality.
[53] This is a very common trick done with real world math problems, especially when two or more models are all valid and there is no way to eliminate any of the remaining models from consideration.

©Brian Gillispie, 2016.

Sometimes, even if the numbers are far apart, the range is still all you have. In those cases, you have to still report the range, and hope that as the time gets closer, you can make a more accurate prediction[54]. So, usually, when these methods are used to make a prediction that ends up giving a wide range for the answer, we usually wait until the desired time is closer, then rerun the model, and update our prediction accordingly then.

---

[54] You can see this in action by looking at a forecast for where a hurricane will be many days into the future. You'll notice a wide range of possible locations, and this is due to the range of answers given by the multiple models.

3.4 Exercises

1. A pond has 12 ducks in the month of May, and 22 ducks in the month of June. How many ducks would be present in the month of July, using the Average Rate of Change?

2. A town has 2598 people in the year 2015, and 2628 in the year 2016. How many people will be present in the year 2017, using the Average Rate of Change?

3. A company sold $3,500 of product on April 20th, and $3,800 of product on April 21st. How much product will they sell on April 23rd, using the Average Rate of Change?

4. An ecologist wishes to predict how many wolves are in a forest. If there are 50 wolves in the month of April, and 48 wolves in the month of May, predict how many wolves will be present in the month of September, using the Average Rate of Change.

5. A town has 3000 people in the year 2015, and 3150 people present in the year 2018. Predict how many people will be present in the year 2020, using the Average Rate of Change.

6. A river has 15000 salmon in the year 2005 and 17000 salmon in the year 2007. Predict how many salmon will be present in the year 2009 using the Average Rate of Change.

7. A college has 2500 students enrolled in 2010, and 2600 students enrolled in 2020. Predict how many students they will have enrolled in the year 2025 using the Average Rate of Change.

8. In the month of May, a sales floor has 250 total sales, and in the month of July a sales floor has 268 total sales. Using the Average Rate of Change, predict how many sales the sales floor will have in the month of October.

©Brian Gillispie, 2016.

9. A pond has 12 ducks in the month of May, and 22 ducks in the month of June. How many ducks would be present in the month of July, using the Geometric Change?

10. A town has 2598 people in the year 2015, and 2628 in the year 2016. How many people will be present in the year 2017, using the Geometric Change?

11. A patient takes a medicine, which releases 325 mg of the medicine in the body. Next hour, there is 275 mg of the medicine in the body. How much medicine will be present 3 hours after the medicine was taken, using the Geometric Change?

12. A deserted island has 100 rabbits on it when first discovered, and next year it is estimated there are 133 rabbits. How many rabbits will be present on the island 5 years after it was discovered, using the Geometric Change?

13. A beehive has 15000 bees in the month of July, and 17000 bees in the month of August. How many bees will be present in the beehive in December, using the Geometric Change?

14*. A sales floor has 250 sales in the month of May and 262 sales in the month of June. Due to a computer error they lost the records saying how many sales they had in April. Use the Geometric Change and work backwards to figure out how many sales they had in April.

©Brian Gillispie, 2016.

For all remaining problems, you are not told which method to use. In these cases, you need to use both methods, and report the range between the two. See Example 3.4.7 for more details.

15. A pond has 12 ducks in the month of May, and 22 ducks in the month of June. Predict how many ducks would be present in the month of July.

16. An author of a bestselling book notices that they sell 17,500 copies in the month of March, and 16,000 copies in the month of April. Estimate how many copies they will sell in the month of May.

17. Repeat Exercise 16, but now estimate how many copies they will sell in the month of August.

18. A polluted pond is in the process of being cleaned up. On September 1st, there are 100 liters of pollutant, and on September 2nd there are 82 liters of pollutant. Estimate how much pollutant will be present on September 4th.

19*. Repeat Exercise 20, but now estimate on what date there will only be 5 Liters of pollutant left.

20*. A patient takes a strong antibiotic. Currently there is 1250 mg of the antibiotic in them, and 3 hours later there is 1000 mg of the antibiotic in them. They cannot take another dose until there is 250 mg of the antibiotic or less in them. Estimate when it is safe to give them another dose. Which answer of the two would you use in this situation, and why?

3.5: Change Tables

So far in this Chapter we have discussed how to make predictions when one or two data points are known. Often times, a modeler will have more than two data points available, and we will need to be able to list the change between all the data points. In an attempt to keep all of this data organized, modelers use what is known as a Change Table, which lists all the data points, then the corresponding change between the data points in one easy to locate location.

To make a Change Table, begin by listing your known points. Put all of the known x values[55] in column one, and all the known y values in column two. Be sure that related points are in the same row, so if one of your data points was $(5, 60)$, then 5 is in column one and 60 in column two of the same row. Then, in column three, list the change between the data points in the current row, and the data points in the previous row. Continue like this until the entire table is filled in.

Let's demonstrate this approach now with a few Examples.

Example 3.5.1: Below is a table listing the number of rabbits in various months. Compute the Average Rate of Change table for this data.

Month	Number of Rabbits
0	46
2	54
4	64
6	76
8	90

Solution to 3.5.1: To calculate the Average Rate of Change table, we need to add on a third column, and in that column we will list the Average Rate of Change for the current row with the previous row. We will abbreviate Average Rate of Change ARC in the table as well. Doing that gives us the following:

---

[55] If your points aren't in (x, y) form, put anything related to time in column one instead.

©Brian Gillispie, 2016.

Month	Number of Rabbits	ARC
0	46	
2	54	
4	64	
6	76	
8	90	

Now, we need to compute the ARC for each row. Since the standard is to list x in column one and y in column two, we will use month for x and y for number of rabbits. This means then in row one, we need to calculate the ARC of $(0, 46)$ with the row before it. Since no row exists before it, we mark the ARC as NA (short for Not Applicable).

Next, we compute the ARC for row two, which means we need to compute the ARC for the points $(0, 46)$ and $(2, 54)$, which is 4. Continuing in like fashion down the table, we calculate the following ARC's for each row which results in a final table of:

Month	Number of Rabbits	ARC
0	46	NA
2	54	4
4	64	5
6	76	6
8	90	7

Example 3.5.2: Below is a table of sales per day for the sales floor at a call center in training. Compute the Arithmetic Change table for the following data.

Day	Number of Sales
Mon	18
Tues	95
Wed	115
Thurs	166
Fri	218

Solution to 3.5.2: Like before, we need to first add on a third column so we can store the Arithmetic Change for each row. Also, since there is no way to compute the Arithmetic Change between rows one and a previous row, we put NA in the first row of that column.

©Brian Gillispie, 2016.

Next, we compute the arithmetic change of each row. The arithmetic change between row two and row one means we need to compute the arithmetic change of the points (Mon, 18), and (Tues, 95), which is 77. Continuing like that through the rest of the table yields the following Arithmetic Change table:

Day	Number of Sales	AC
Mon	18	NA
Tues	95	77
Wed	115	20
Thurs	166	51
Fri	218	52

Example 3.5.3: The following table lists the population of bats in a local cave for various months. Compute the Geometric Change table for this data.

Month	Number of Bats
5	600
6	700
7	775
8	825
9	890

Solution to 3.5.3: Just like the previous Examples, we start by adding a column for the Geometric Change to the table (abbreviated GC), and we put NA in the first row of that column.

Next, we calculate the Geometric Change for each row with the previous row. For row two, that means we need to calculate the Geometric Change between the points $(5, 600)$ and $(6, 700)$. If we do that, we end up with a Geometric Change of 1.17, rounded to the nearest hundredth. Continuing in that fashion yields us the following table:

Month	Number of Bats	GC
5	600	NA
6	700	1.17
7	775	1.11
8	825	1.06
9	890	1.08

©Brian Gillispie, 2016.

3.5 Exercises

Calculate the appropriate change table for each of the data sets provided.

1. Calculate the Average Rate of Change table for the following data:

Month	Number of Pigeons
0	25
3	43
6	67
9	79
12	82

2. Calculate the Average Rate of Change table for the following data:

Seconds	Height of Ball
0	1500
5	500
10	0

3. Calculate the Average Rate of Change table for the following data:

Hour	Cells
0	500
1	550
2	650
10	2650
11	2675

4. Calculate the Arithmetic Change table for the following data.

Year	Number of Rabbits
0	12
1	30
2	75
3	188
4	470

5. For the data in Exercise 4, calculate the Geometric Change table for the following data.

6. Calculate the Arithmetic Change table for the following data.

Day	Burgers Sold
Monday	985
Tuesday	1012
Wednesday	500
Thursday	750
Friday	1125

7. Calculate the Arithmetic Change table for the following data.

Hour	Gnats
0	752
1	742
2	532
3	258
4	12
5	0

8. For the data in Exercise 7, calculate the Geometric Change table for the following data.

9. Calculate the Arithmetic Change table for the following data.

Seconds	Speed (in mph)
0	5
1	12
2	20
3	30
4	35
5	41

10. For the data in Exercise 9, calculate the Geometric Change table.

11*. Calculate the Arithmetic Change table for the following data. **Warning:** *The data in this table is not in sequential order. Rescale the data so that the x values are sequential, then calculate the Arithmetic Change.*

Month	Ducks
5	75
8	88
11	102
14	125
17	144
20	160

12* For the data in Exercise 11, in what units would the Arithmetic Change be in?

13*. For the data in Exercise 11, calculate the Geometric Change table.

©Brian Gillispie, 2016.

3.6: Using Change: Three Point Modeling, Sequential Points

So far in this Chapter we have worked with making predictions when only one or two data points are provided. In this Section, we are going to discuss what to do when we are given three data points. As it is unlikely that the Average Rate of Change or the Geometric Change will be constant between all the data points, we will need to account for the change in whichever method we use to measure change as well when making our prediction.

In this Section we will discuss the special case of when three points are provided, but the three points are set up in such a way that the x values are sequential. This will allow us to more easily measure and calculate the change in the change than if the points were not sequential, as we will be able to use the Arithmetic Change or the Geometric Change now.

3.6.1: Change of the Change: Idea and Definition

To begin our discussion of modeling with three data points, we need some notation, as it will become cumbersome to say Change of the Arithmetic Change or Change of the Geometric Change all the time. Therefore, we will use the mathematical standard of putting the symbol delta (Δ) in front of the variable that is changing. Therefore, this means that if you see the notation ΔAC it will mean the Change of the Arithmetic Change, and if you see the notation ΔGC it will mean Change of the Geometric Change.

With the notation now defined, let's discuss how to calculate Change in the Arithmetic Change. By definition, change between two points is defined as the difference in the value divided by the difference in time. That means we need to compute:

$$\Delta AC = \frac{\text{Change in Arithmetic Change}}{\text{Change in Time}}$$

However, since the Arithmetic Change itself changes over one time unit, therefore, since the points in our data set are sequential, this means that our change in time is one. This means the definition of ΔAC is the total change between the two Arithmetic Changes, and is defined as follows:

> **Definition 3.6.1:** The Arithmetic Change of the Arithmetic Change (denoted ΔAC) is how much the Arithmetic Change is itself changing. Specifically, if we have three points (x_1, y_1), (x_2, y_2) and (x_3, y_3), which are all equally spaced, then the Arithmetic Change of the Arithmetic Change is computed as follows:
>
> $$\Delta AC = (y_3 - y_2) - (y_2 - y_1)$$

Notice what is going on here. The first term in the ΔAC is $(y_3 - y_2)$, which is how the Arithmetic Change of the second and third point is calculated. Then, we are subtracting that term by $(y_2 - y_1)$, which is the Arithmetic Change of the first and second point. So, you can compute the ΔAC by using the formula above, or by calculating the Arithmetic Change first, then subtracting the Arithmetic Change of the third and second point from the Arithmetic Change of the second and first point.

Let's now use this definition to calculate the ΔAC for a few Examples before proceeding:

Example 3.6.1: For the data points $(0, 50)$, $(1, 60)$, and $(2, 75)$, what is ΔAC?

Solution to 3.6.1: To compute the ΔAC, we need to first find the Arithmetic Change between each of the data points. The Arithmetic Change of the first and second point is 10, and the Arithmetic Change of the second and third point is 15. With those found, we then compute ΔAC by computing the difference between the Arithmetic Changes[56], which we compute as:

$$\Delta AC = 15 - 10 = 5$$

Therefore, the ΔAC for these three data points is 5.

---

[56] It is important to subtract the values in the right order. Always subtract the arithmetic change between the second and third point from the arithmetic change between the first and second point.

©Brian Gillispie, 2016.

Since sometimes the data is provided not in point form, let's work out an example where the data is provided in a different way next.

Example 3.6.2: It is 62 degrees outside at 1 am, 57 degrees outside at 2 am, and 55 degrees outside at 3 am. Compute ΔAC for these temperatures.

Solution to 3.6.2: This time, we need to put the data into point form before beginning. Since time is x, we will let x be the time in hours after midnight[57], which means the associated temperature will then be y. With that decided on, that makes our three points $(1, 62)$, $(2, 57)$, and $(3, 55)$.

Next, find the Arithmetic Change between the each of the data points. The Arithmetic Change between the first and second point is -5, and the Arithmetic Change between the second and third point is -2. Therefore, ΔAC is:

$$\Delta AC = -2 - (-5) = 3$$

Therefore, the ΔAC for these points is 3.

We can define ΔGC similarly, by treating it as a rate of change divided by time. However, since ΔGC is very rarely used in practice, we will not define it at this point in time.

---

[57] It can be hours after any time actually. All that matters is the points stay sequentially spaced.

©Brian Gillispie, 2016.

3.6.2: Using the Change of the Change

Now that we have defined the Change of the Change, we will discuss how to use this to make a prediction given three points. To begin, let's consider a hypothetical problem. Pretend we know that in the month of May there are 46 rabbits, and in the month of June there are 50 rabbits, and in the month of July there are 55 rabbits. With this information, and x defined as month and y defined as number of rabbits, our data points are $(5, 46)$, $(6, 50)$, and $(7, 55)$. Also, note that our Arithmetic Change between points one and two is 4 and the Arithmetic Change between points two and three is 5, and the corresponding ΔAC is 1. This can all be represented in table form as the following:

Month	Rabbits	Arithmetic Change	ΔAC
5	46	N/A	N/A
6	50	4	N/A
7	55	5	1

Now, let's say we want to predict how many rabbits there will be in the month of August, or month 8. Previously, when we knew the Arithmetic Change, we would take the previous value (55 in this case) and add the most recent Arithmetic Change to it (5 in this case) to predict the new value. However, if we do that, we will be ignoring the fact that the Arithmetic Change is itself changing! To compensate for that, we first add the ΔAC to the Arithmetic Change, then add the new Arithmetic Change to the previously known value. This means that we would then say that the number of rabbits present in August is July's Rabbits + (Arithmetic Change between June and July + ΔAC between the last three points). Doing this means that for this data, we would predict that in the month of August there will be $55 + (5 + 1)$ rabbits, or, 61 rabbits in the month of August.

Now, what if we want to predict the number of rabbits in September instead? First, we would need to calculate the number of rabbits present in August, and update our table with what we computed. If we do that, the table now looks like the following:

Month	Rabbits	Arithmetic Change	ΔAC
5	46	N/A	N/A
6	50	4	N/A
7	55	5	1
8	61	6	1

Notice that in the ΔAC column we carried down the previous ΔAC. This is because when there are three points, we assume that the ΔAC remains constant all throughout[58]. Therefore, this means that once we know that ΔAC is one, then ΔAC remains 1 all throughout the problem. Also notice that for the Arithmetic Change in the row for August we put 6. This is because the Arithmetic Change from June to July needs to increase by 1 per the calculated ΔAC. Therefore, we take June to July's Arithmetic Change, add ΔAC to it then put that as the result in the August row (as this is July to August's Arithmetic's Change). Or, an easier way to think of this is in to update your Arithmetic Change, add the number in the ΔAC column that is diagonal from it, then put the result in the row below[59]. Alternatively, you can think of it as an algebra equation where you need to find what Arithmetic Change I would have to have such that if I subtract the current Arithmetic Change, I get this ΔAC. Either method works.

Now that we have updated our table, we want to compute the number of rabbits present in September. To do that, we need to take the number of rabbits in August (61), add to it the updated Arithmetic Change (6), then add to that the ΔAC (1). Doing that gives us the following:

$$61 + 6 + 1$$

Which means that in September there will be 68 rabbits then, based on our table.

As you can see, this is a lot to keep track of. We highly recommend using a Change Table to store all the known information, and update it as needed. We'll do that in our next Example.

---

[58] We would need four known points to calculate if the ΔAC itself was changing. But by then the techniques being used are becoming quite complex and cumbersome and other approaches are recommended instead.

[59] You can also use this method to update the number of rabbits. Take July's number of rabbits (55), add to it the Arithmetic Change down and diagonal from it (6) and put the result in the row below. You'll get 61.

Example 3.6.3: An ecologist has noticed that in a forest there were 28 kangaroos in the month of February, 32 kangaroos in the month of March, and 35 kangaroos in the month of April. Using the Arithmetic Change and the ΔAC, predict how many kangaroos will be present in the month of July.

Solution to 3.6.3: First, we need to figure out what our points are. Let's define x as the month, and y as the number of kangaroos. Then, we end up with our points being $(2, 28)$, $(3, 32)$ and $(4, 35)$. Calculating the resulting Arithmetic Change and ΔAC, and putting those in table form results in the following table:

Month	Kangaroos	Arithmetic Change	ΔAC
2	28	N/A	N/A
3	32	4	N/A
4	35	3	-1

Next, we need to work our way down the table. First, carry down the ΔAC, it will be the same for any rows we add to the table. Add on the rows for May, June and July, and put our value for ΔAC in those rows. This will result in the following table:

Month	Kangaroos	Arithmetic Change	ΔAC
2	28	N/A	N/A
3	32	4	N/A
4	35	3	-1
5			-1
6			-1
7			-1

Next, we need to update our Arithmetic Change all the way down the table. The easiest way to do this is to take the current known Arithmetic Change, add it to the value for ΔAC down and diagonal from it, and put the new result below the current Arithmetic Change. If we do that we get that in the row for Month 5, our Arithmetic Change will be $3 + (-1)$, or 2. In the row for Month 6, our Arithmetic Change will then be $2 + (-1)$, or 1. Continuing like this will result in the following table with the Arithmetic Change column all filled in:

Month	Kangaroos	Arithmetic Change	ΔAC
2	28	N/A	N/A
3	32	4	N/A
4	35	3	−1
5		2	−1
6		1	−1
7		0	−1

To finish, we need to update the number of kangaroos all the way down the table until we reach our desired month. The easiest way to do this is to take the currently known number of kangaroos, add to it the Arithmetic Change down and diagonal from it, then put the new result below. If we do that, we get that in the row for Month 5, our number of kangaroos will be 35 + 2, or 37. Continuing in similar fashion all the way down the table results in the following final table:

Month	Kangaroos	Arithmetic Change	ΔAC
2	28	N/A	N/A
3	32	4	N/A
4	35	3	−1
5	37	2	−1
6	38	1	−1
7	38	0	−1

Therefore, we conclude that in the month of July, there will be 38 kangaroos present.

Let's work out one more Example now to close out this Section. In this one, we will explain how to find the values on the table in a slightly different way.

Example 3.6.4: A wildlife reserve notices that there are 1700 birds present in the month of July, 1670 birds present in the month of August, and 1660 birds present in the month of September. Using the Arithmetic Change and the ΔAC, predict how many birds will be present in next January?

3.6: Using Change: Three Point Modeling, Sequential Points

Solution to 3.6.4: Like last time, we need to start out by figuring out what our points are. Using the rules from Chapter 2, we will let x be months and y be birds, and we will set up our data points to be $(7, 1700)$, $(8, 1670)$, and $(9, 1660)$. We then calculate the Arithmetic Change between the points and the ΔAC, and place our results in a Change Table as such:

Month	Birds	Arithmetic Change	ΔAC
7	1700	N/A	N/A
8	1670	−30	N/A
9	1660	−10	20

Next, we expand the table, and carry down the ΔAC. Since we need to calculate the number of birds in next January, we will expand the table down to month 13. Doing so gives us the following table:

Month	Birds	Arithmetic Change	ΔAC
7	1700	N/A	N/A
8	1670	−30	N/A
9	1660	−10	20
10			20
11			20
12			20
13			20

Next, calculate the Arithmetic Change for each row. This time, we are going to find the values differently than we did in Example 3.6.3. This time, the way we will find our Arithmetic Change is we will think of it as an algebra problem where we need to figure out what number can I subtract my currently known Arithmetic Change from and get the ΔAC value we have. Or, in other words, I need to solve the following to find the Arithmetic Change for Month 10:

$$x - (-10) = 20$$

Where x is the Arithmetic Change I want to find, and I want it to be such that if I subtract my most recent Arithmetic Change of -10 I end up with 20. If I solve that equation, I find that x is 10, so my missing Arithmetic Change is 10 in the row for Month 10.

©Brian Gillispie, 2016.

Now, to find the Arithmetic Change for Month 11, I need find what number can I subtract 10 from (Month 10's Arithmetic Change) to get my ΔAC of 20. This means I need to solve the equation:

$$x - 10 = 20$$

Which means my missing Arithmetic Change is 30 for Month 11. Continuing like this down the table, I end up with the following for my Arithmetic Changes:

Month	Birds	Arithmetic Change	ΔAC
7	1700	N/A	N/A
8	1670	−30	N/A
9	1660	−10	20
10		10	20
11		30	20
12		50	20
13		70	20

Now, to find the number of birds for each month, we will solve a similar equation. The number of birds in Month 10 needs to be such that if I subtract the number of birds in Month 9, I'll end up with the Arithmetic Change listed in Month 10's row. In other words, I need to solve the following equation:

$$x - 1660 = 10$$

Which, if we solve that, we get that x is 1670, or, there should 1670 birds in Month 10. Now, we can find the number of birds in Month 11 similarly, as the number of birds in Month 11 minus the number of birds in Month 10 must equal 30, or the Arithmetic Change in the row for Month 11. This means we have to solve the following:

$$x - 1670 = 30$$

Which, once we solve that equation, we get that the number of birds in Month 11 is 1700. Continuing like this gets us the following for our final table:

Month	Kangaroos	Arithmetic Change	ΔAC
7	1700	N/A	N/A
8	1670	-30	N/A
9	1660	-10	20
10	1670	10	20
11	1700	30	20
12	1750	50	20
13	1820	70	20

Or, in January of next year, there should be 1820 birds at the wildlife refuge.

3.6 Exercises

1. In the month of January, 32 owls are present in a forest. In February, 33 owls are present. In March, 35 owls are present. Use the arithmetic change and the ΔAC to predict how many owls will be present in April.

2. For the data in Exercise 1, predict how many owls will be present in May.

3. For the data in Exercise 1, predict how many owls will be present in June.

4. In 2015 there are 15 goats on an island. In 2016 there are 19 goats on an island. In 2017, there are 24 goats on an island. Use the arithmetic change and the ΔAC to predict how many goats will be present in 2018.

5. For the data in Exercise 4, predict how many goats will be present in 2019.

6. For the data in Exercise 4, predict how many goats will be present in 2024.

7. A scientist is interested to see how fast a chemical can remove cancer cells. If a patient has a tumor that weighs 20 g when the chemical is administered at 2 pm, then the tumor weighs 19.5 g at 3 pm, and 19.1 g at 4 pm, predict how much the tumor will weigh at 5 pm using the Arithmetic Change and the ΔAC.

8. For the data in Exercise 7, predict how much the tumor will weight at 8 pm.

9. For the data in Exercise 7, figure out when the tumor stops shrinking. *Hint: Calculate the weight of the tumor until you get two hours that the tumor weighs the same amount.*

10. In a wildlife reservoir, there were 250 ducks in the year 2010, 260 ducks in the year 2011, and 274 ducks in the year 2012. Predict how many ducks will be present in the year 2013 using the Arithmetic Change and the ΔAC.

©Brian Gillispie, 2016.

11. For the data in Exercise 10, predict how many ducks will be present in the year 2014.

12*. For the data in Exercise 10, calculate what would be a reasonable number of ducks present back in 2009, based on this data.

13. A lake has been polluted! On April 26th there are 5 liters of pollutant, on April 27th there are 5.7 liters of pollutant, and on April 28th there are 5.9 liters of pollutant. Use the Arithmetic Change and the ΔAC to predict how much pollutant will be present on April 29th.

14. For the data in Exercise 13, predict how much pollutant will be present on May 2nd.

15. For the data in Exercise 13, predict on what date the pollutant will be gone.

16*. For the data in Exercise 13, predict on what date the pollutant was first introduced into the water.

17*. In a wildlife reservoir, there were 250 ducks in the year 2010, 260 ducks in the year 2011, and 274 ducks in the year 2012. Predict how many ducks will be present in the year 2013 using the Geometric Change and the ΔGC. *Hint: Think of a reasonable way to define ΔGC based on the ideas of this section.*

©Brian Gillispie, 2016.

Chapter 3 Summary

3.1: In this Section, we defined the various ways to calculate change. Given two points (x_1, y_1) and (x_2, y_2), we defined the Total Change (TC) as:

$$TC = y_2 - y_1$$

The Average Rate of Change was defined as:

$$ARC = \frac{y_2 - y_1}{x_2 - x_1}$$

If the points are sequential, then the Arithmetic Change is defined as:

$$AC = y_2 - y_1$$

And, the geometric Change is defined as:

$$Geometric\ Change = \frac{y_2}{y_1}$$

3.2: In this Section we discussed how the Average Rate of Change can be used to tell the rate of increase or decrease between two numbers, and how the Geometric Change can be used to calculate rates of increase or decrease.

©Brian Gillispie, 2016.

3.3: In this Section we discussed how we can use the Arithmetic Change and the Geometric Change to predict a future value if we know one point and either the rate of change, or the percent increase or decrease. Specifically, if we know the Arithmetic Change, the next value can be estimated by:

$$New\ value = Old\ Value + Arithmetic\ Change$$

Similarly, if we know the Geometric Change, we can estimate the new value by:

$$New\ value = Old\ Value * Geometric\ Change$$

3.4: In this Section we discussed how we can make predictions if we have two data points. If we have two data points, we use those two points to calculate the Arithmetic Change or the Geometric Change, then calculate the next value by the methods of Section 3.3.

3.5: In this Section we discussed how to create and display a change table. The change in a row is always the change between the points in that row and the points in the previous row, and if there is no previous row we put NA.

3.6: In this Section we discussed how to use three sequential data points to predict future values by calculating the Arithmetic Change and the Change of the Arithmetic Change (ΔAC). In addition, we discussed how to use these to predict future values via a table.

©Brian Gillispie, 2016.

©Brian Gillispie, 2016.

Chapter 4: Discrete Modeling with Sequences

One of the more common models used in modeling is the discrete sequence. In this Chapter, we will discuss how to define a sequence, as well as how to set up recursive sequences that model a given situation. Then, we will explore how to find and use fixed points of recursive sequences to make predictions about the long term behavior of the model. And finally, we will explore how to use sequences that are interconnected with each other, and see how to use these to model the spread of a disease in an area.

©Brian Gillispie, 2016.

©Brian Gillispie, 2016.

4.1: Sequences

In order to begin our discussion of discrete sequences, we need to define what sequences are, as well as the notation needed to work with sequences. In this Section, we will cover what a sequence is, as well as the sequence notation used in modeling.

4.1.1: What is a Sequence?

A sequence is defined as follows:

> *Definition 4.1.1:* A **sequence** is an ordered list of objects or numbers

Based on this definition, a sequence can be anything from a list of the American Presidents, in the order they were in office, to a list of the integers in order. However, notice that in the definition, a sequence is an **ordered** list. Therefore, if we had the sequence:

$$1, 2, 3, 4, 5, 6, 7, 8, 9, 10, \ldots$$

This sequence is a different sequence than the sequence:

$$10, 9, 8, 7, 6, 5, 4, 3, 2, 1, \ldots$$

Even though both sequences list the same numbers, because the numbers are not in the same order, the two sequences are not considered to be the same sequence. In sequences, the order the items in the sequence are listed does matter. For that reason, it is important to always list the items in a sequence in the proper order.

©Brian Gillispie, 2016.

4.1.2: Sequence Notation

While a modeler could always define sequences by listing out every single item in the sequence, this gets cumbersome very quickly. For example, one does not want to list this sequence by hand every time:

$$1, 2, 4, 8, 16, 32, 64, 128, 256, 512, 1024, 2048, 4096, 8192, \ldots$$

Because we don't want to have to list out this sequence every time we reference it, we need a notation to reference a sequence and elements in a sequence. In addition, we would like an easy way to reference individual elements within the sequence. With sequences, the standard is to use a capital letter to denote the entire sequence, then, reference individual elements with the following notation:

> *Definition 4.1.2:* The symbol a_n is used to refer to the nth element of the ordered sequence represented by the letter A

The choice of the letter A is arbitrary, any letter can be used to represent the sequence that the modeler wishes to use. However, it is advised to use a letter that is easy to read and not confused with the subscript beside it when referencing elements. For that reason, A and B are the most common letters used for sequences.

Let's demonstrate this notation now with a few Examples:

Example 4.1.1: Given the sequence $A = \{1, 2, 4, 8, 16, \ldots\}$, what is a_3?

Solution to 4.1.1: a_3 refers to the third element of the sequence listed. The third element is 4, so we would conclude that $a_3 = 4$

Example 4.1.2: Given the sequence $A = \{1, 1, 2, 3, 4, 8, 13, 21, ...\}$, what is a_6?

Solution to 4.1.2: a_6 refers to the sixth element of the given sequence. The sixth element is 8, so we would conclude that $a_6 = 8$

In both of the previous Examples, the sequence was listed out explicitly. This is not common practice, as it is still cumbersome to have to write out the entire sequence so we can reference it. Instead, it is common to provide a rule that the modeler can use to figure out each element in the sequence as they are needed. These rules often look like:

$$a_n = some\ rule\ involving\ n$$

Sequences defined this way follow the same rules as functions do. Let's demonstrate with a few Examples:

Example 4.1.3: Given the sequence $a_n = 2n + 1$, what is a_5?

Solution to 4.1.3: This time we are given a formula by which we may find a_5. To find this term, plug in 5 for all the n's. Doing this yields:

$$a_5 = 2(5) + 1 = 11$$

Or, the term a_5 of the sequence is 11.

Example 4.1.4: Given the sequence $a_n = 2n^2 + n$, what is a_6?

Solution to 4.1.4: Like the last example, we are given a formula to find the desired term. To find our desired term, plug in 6 for all the n's in the formula. Doing this yields:

$$a_6 = 2 * 6^2 + 6 = 78$$

©Brian Gillispie, 2016.

Or, the term a_6 of the sequence is 78.

Example 4.1.5: Given the sequence $a_n = 100 * (0.5)^n$, what is a_{10}?

Solution to 4.1.5: Again, we are given a formula to find a desired term. This time, we wish to know what the tenth term in the sequence is. To find this term, plug in 10 for all the n's in the formula. If we do this, we get:

$$a_{10} = 100 * (0.5)^{10} \approx 0.0977$$

Or, the tenth term in the sequence is about 0.0977.

Defining sequences in terms of a formula allows the modeler to easily calculate what the desired term in the sequence is, as you can see.

©Brian Gillispie, 2016.

4.1 Exercises

Given the sequence $A = \{5, 9, 13, 17, 21, 25, 29, 33, 37, 41, 45, 49, ...\}$, find the following:

1. a_3?

2. a_7?

3. a_{13}?

4. a_{10}?

5* a_{20}?

Find the desired value for the given sequence.

6. $a_n = 2^n$, what is a_5?

7. $a_n = 4n - 8$, what is a_6?

8. $a_n = 2n + 6$, what is a_3?

9. $a_n = n^2$, what is a_5?

10. $a_n = 100 * (0.75)^n$, what is a_5?

11. $a_n = 500 * (1.05)^n$, what is a_4?

12. $a_n = n^2 - n + 6$, what is a_0?

©Brian Gillispie, 2016.

4.2: Recursive Sequences

In the previous Section we discussed what sequences are, as well as the sequence notation that is often used in modeling. In this Section, we will cover another way to represent a sequence via a recursive sequence.

4.2.1: Recursive Sequence Representation

A recursive sequence is a sequence where the formula for the current value in the sequence depends on the previous values in the sequence. An example of a recursive sequence is:

$$a_n = 2 * a_{n-1}$$

This example is saying that the current value, denoted by a_n, is equal to two times the previous value, which is denoted by a_{n-1}. Any time you see the term a_{n-1} in a recursive sequence definition, know that this stands for the value before the current value.

Let's now demonstrate this in a couple of Examples.

Example 4.2.1: Explain what the sequence $a_n = 1.04 * a_{n-1}$ means

Solution to 4.2.1: In this sequence, a_{n-1} means the previous term, so this means we are to take 1.04 times the previous term to acquire the new term in the sequence.

Example 4.2.2: Explain what the sequence $a_n = a_{n-1} + 5$ means

Solution to 4.2.2: In this sequence, a_{n-1} means the previous term, so this means that we are to take the previous term and add 5 to it to acquire the new term in the sequence.

©Brian Gillispie, 2016.

4.2.2: Calculating terms in a Recursive Sequence

Now that we have discussed what a recursive sequence is, let's now discuss how to use them to calculate terms in a sequence. Let's start this discussion by considering the sequence $a_n = 1.04 * a_{n-1}$, which we covered in Example 4.2.1. As we saw in Example 4.2.1, this means that the current value of the sequence is equal to 1.04 times the previous value in the sequence. Therefore, if we wished to compute what a_7 is, we have the following:

$$a_7 = 1.04 * a_6$$

Notice we replaced a_{n-1} with a_6, as a_6 is the previous value in the sequence for a_7. However, this creates a new problem, as we don't know what a_6 is! No problem, let's use the definition of the sequence to find a_6. If we do that, we end up with:

$$a_6 = 1.04 * a_5$$

However, we don't know what a_5 is either. We could try to find a_5 by using the sequence definition, but then we would need to find the value of the term a_4, which would then require the value of the term a_3, as so on. This will never end, as each term we try to compute will require us to learn the value of the previous term. Therefore, with the information we have so far, it is impossible for us to ever compute the value of any one term of the sequence.

It turns out we can easily fix this problem by providing more information. Modelers prevent this problem from occurring by providing what is called an initial condition of the sequence, denoted a_0 or a_1, depending on how the sequence is to be created[60]. With this initial condition, it is possible to calculate the values of other terms in the sequence, which we will demonstrate now.

---

[60] Which one depends on many factors. However, it is common to use a_0 if time is involved, so as to let the initial condition represent starting time.

©Brian Gillispie, 2016.

Example 4.2.3: Given the sequence $a_n = 1.04 * a_{n-1}$, and initial condition $a_0 = 100$, find a_1

Solution to 4.2.3: In this Example, we wish compute the value of a_1, and we know that a_1 is 1.04 times the previous value, or:

$$a_1 = 1.04 * a_0$$

Since we were given that a_0 is 100, plug that into the equation above:

$$a_1 = 1.04 * 100$$

Which, once we complete the calculations, gives us $a_1 = 104$

Example 4.2.4: Given the sequence $a_n = 0.9 * a_{n-1}$, and the initial condition $a_0 = 50$, find a_3

Solution to 4.2.4: In this example, we want to calculate the value of a_3, and we know that a_3 is 0.9 times the previous value, or:

$$a_3 = 0.9 * a_2$$

However, we don't know the value of the second term either, as the second term is 0.9 times the previous term, or:

$$a_2 = 0.9 * a_1$$

Again, we need the value of a term we don't have, the value of the first term. Since the first term is 0.9 times the previous value, this gives us:

$$a_1 = 0.9 * a_0$$

©Brian Gillispie, 2016.

We know what a_0 is though, as that was provided as an initial condition. Plugging in our given value for a_0 into the formula for a_1 gives us:

$$a_1 = 0.9 * 50 = 45$$

Since we now know the value of the term a_1, plug that into our formula to a_2, which gives us:

$$a_2 = 0.9 * 45 = 40.5$$

With a_2 known, we can then plug that into our formula for a_3, which gives us:

$$a_3 = 0.9 * 40.5 = 36.45$$

Therefore, the third term of the sequence (a_3) is 36.45.

The reader has probably noticed that in the last Example, it took a lot of work to find the value of the third term of the sequence. In general, finding the value of a term that is deep in a recursive sequence takes a long time, as you have to work forwards from the given starting value until you reach your desired term, calculating the value of all of the terms along the way. For example, if we were given the value of a_0, and wanted to find the value of the term a_{25}, we would need to compute the value of a_1, then a_2, then a_3, and so on all the way up to the desired term. This can take a long time, so for that reason modelers usually try to solve the recursive sequence into an explicit formula (like those in Section 4.1), or use technology to calculate the values of sequences.

Recursive sequences can also have dependency on n as well as the previous value, as this next Example will demonstrate.

©Brian Gillispie, 2016.

Example 4.2.5: Given the sequence $a_n = 0.6 * a_{n-1} + 2n$, and the initial condition $a_0 = 25$, find a_2

Solution to 4.2.5: Since the recursive sequence starts at a_0, we will use it to first find a_1, then a_2. Plugging into our recursive formula for a_1, we end up with:

$$a_1 = 0.6 * 25 + 2(1)$$

Notice the n in the $2n$ term is replaced by 1, because we are trying to calculate the term a_1 of the sequence. In general, any time you see an n in the formula on the right hand side of the equal sign, you replace it by the number of the sequence you are trying to compute (which is 1 in this case). Calculating this out gives us:

$$a_1 = 17$$

Now, plugging that into our formula for a_2 gives us:

$$a_2 = 0.6 * 17 + 2 * 2 = 14.2$$

Therefore, the second term of this sequence (a_2) is 14.2.

4.2.3: Excel Representation of Recursive Sequences

Since technology is highly useful for representing recursive sequences, in this Section we will list how a modeler can use Excel 2010 to represent and calculate values in a recursive sequence. The directions should still work for other versions of excel, but unfortunately we cannot guarantee this, as technology is constantly changing.

In this section, we will show how we can use excel to find the value a_{25} for the sequence $a_n = 1.04 * a_{n-1}$, with the initial value $a_0 = 100$. To begin, open up a new spreadsheet. When you do that, your screen should look like this:

©Brian Gillispie, 2016.

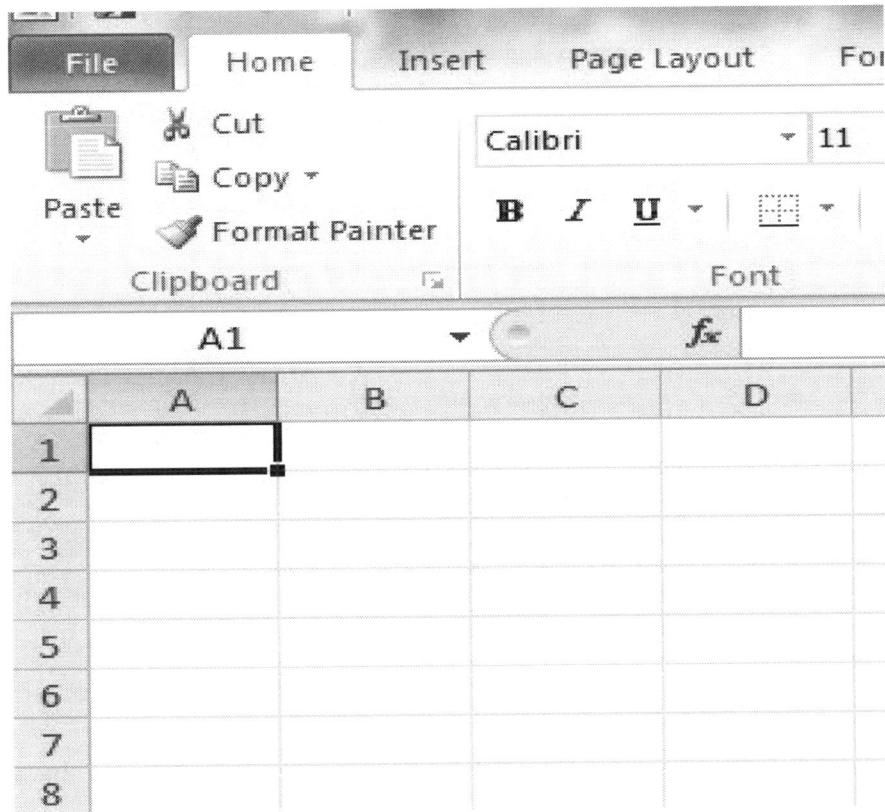

Notice how the columns in excel are labeled A, B, C, D, and the rows are labeled 1, 2, 3, 4, etc. A cell in excel is often referred to by its row and column, so if we say put things in cell A2, we mean Column A, Row 2.

To represent our recursive sequence in excel, put in the cell A1 the initial condition for the sequence. So for this example, enter into the cell A1 the value 100, which is our initial condition. Next, in cell A2, enter in the formula for the recursive sequence, using the fact that the starting value is currently in A1. In our example, the sequence is defined as 1.04 times the previous value. To enter that into excel, enter the following: =1.04 * A1. The equal sign tells excel that we are entering a formula, and A1 tells excel to use the value currently in cell A1. Having completed that, excel will look like the following:

	A	B	C	D
1	100			
2	104			
3				
4				
5				

With the starting value entered in cell A1, and the formula for the sequence entered into cell A2, we are now ready to have excel calculate the rest of the values of the sequence for us. In order to do this, move your cursor in excel to the lower right hand corner of the cell A2. It should turn to a black cross. If the cursor is still a solid, white, cross, then the cursor is not in the correct spot yet. Once the cursor has turned to a smaller, black cross, click and hold down the left mouse button. Then, drag the cursor down to your desired cell, and release. In this example, since we want to calculate the value a_{25}, we need to go down to cell 26, since cell one was used for the initial value, which was a_0.

Once all of this has been done, all the modeler needs to do is go down to the appropriate row and read off the value, remembering that we put the initial value in row 1. So for our example, we need the value in cell A26, which is row 26, column A. If we look at the value there, we will find it is 268.5836. Therefore, this means that $a_{25} = 268.5836$ for our Example[61].

---

[61] Be very careful! The rows will NOT line up with the subscript for the sequence. This is because sequences usually start at 0, but excel starts at row 1.

©Brian Gillispie, 2016.

4.2 Exercises

Given the following recursive sequence and starting value, find the asked for value.

1. $a_n = 2 * a_{n-1}$, initial value is $a_0 = 3$, find a_2

2. $a_n = 500 + a_{n-1}$, initial value is $a_0 = 15$, find a_3

3. $a_n = 0.96 * a_{n-1}$, initial value is $a_0 = 250$, find a_1

4. $a_n = a_{n-1} + 600$, initial value is $a_0 = 1000$, find a_4

5. $a_n = 1.004 * a_{n-1} - 1000$, initial value is $a_0 = 90{,}000$, find a_1

6. $a_n = 0.95 * a_{n-1} + 300$, initial value is $a_0 = 600$, find a_3

7. $a_n = 1.02 * a_{n-1}$, initial value is $a_0 = 500$, find a_3

8. $a_n = 1.005 a_{n-1} - 2n$, initial value is $a_0 = 100$, find a_1

9. $a_n = a_{n-1} + n + 3$, initial value is $a_0 = 46$, find a_4

10. $a_n = 0.97 a_{n-1} + 0.25 n$, initial value is $a_0 = 100$, find a_3

11. $a_n = 1.2 a_{n-1} - 10n$, initial value is $a_0 = 1000$, find a_2

12. $a_n = a_{n-1} - 5n$, initial value is $a_0 = 1000$, find a_4

©Brian Gillispie, 2016.

13* $a_n = 1.2a_{n-1} - 0.7a_{n-2}$, initial values are $a_0 = 50$, $a_1 = 40$, find a_2

14* $a_n = a_{n-1} + a_{n-2}$, initial values are $a_0 = 1$, $a_1 = 1$, find a_5

15* $a_n = 1.05a_{n-3} + a_{n-1}$, initial values are $a_0 = 1000$, $a_1 = 1000$, $a_2 = 2000$ find a_4

16. $a_n = 1.004 * a_{n-1} - 1000$, initial value is $a_0 = 90{,}000$, use excel to find a_{40}

17. $a_n = 0.95 * a_{n-1} + 300$, initial value is $a_0 = 600$. Use excel to find a_{25}

18. $a_n = 1.2 * a_{n-1}$, initial value is $a_0 = 100$. Use excel to find a_{12}

19. $a_n = 0.99a_{n-1} - 500$, initial value is $a_0 = 500{,}000$. Use excel to find a_{100}

20. $a_n = a_{n-1} - 500$, initial value is $a_0 = 100{,}000$. Use excel to find a_{25}

21*. $a_n = a_{n-1} + n + 3$, initial value is $a_0 = 46$. Use excel to find a_{13}

22* $a_n = 1.2a_{n-1} - 0.7a_{n-2}$, initial values are $a_0 = 50$, $a_1 = 40$. Use excel to find a_{10}

©Brian Gillispie, 2016.

4.3: Creating Recursive Sequences

Now that we know and understand how to use sequences and recursive sequences, we are ready to learn how to convert a sequence or a set of data to a recursive sequence. We will begin by looking at how to create a sequence that is valid for two given data points, then will move on to how to create a sequence if we know the starting value and one of the rates of change.

4.3.1: Sequences from Two Points

Previously, when we had two data points, we used them and the Average Rate of Change or the Geometric Change to make predictions of future values. Now, we wish to use those two points as well as the Rates of Change to create a sequence. To do this, let's first look closely at what we did back in Chapter 3 when we wished to make a prediction. For simplicity, we will consider the Geometric Change case first.

Assume we have two points which represent the first two values of our sequence. How can we use them to generate a formula that we can use to predict future points? The easiest way to do this is look at how we came up with those future points originally. Back in Chapter 3, we said that if we know two points, we could calculate the Geometric Change, then, take one of the data points times the Geometric Change to find the next data point. In other words, if we know the two values of our sequence are a_0 and a_1, then we found a_2 by computing:

$$a_2 = a_1 * GC$$

Similarly, if we wanted a_3, then we computed:

$$a_3 = a_2 * GC$$

In both of these cases, the value we wanted to compute is dependent on the previous value. This means we could then instead write our calculations as a sequence as follows:

©Brian Gillispie, 2016.

$$a_n = a_{n-1} * GC$$

Which would mean the same thing, take the previous value times the Geometric Change to get the next value. This means in general, if we have two points, one way we can create a recursive sequence is to compute the Geometric Change, then set up our sequence as:

$$a_n = a_{n-1} * GC$$

Similar[62] reasoning for the Average Rate of Change yields that another valid recursive sequence is:

$$a_n = a_{n-1} + ARC$$

Just remember to state your starting value too, else both recursive definitions will be nonsense. Now, let's demonstrate finding these formulas with a few Examples.

Example 4.3.1: Given the first two terms of the sequence $a_0 = 20$, $a_1 = 30$, create a recursive sequence for these points using the Geometric Change.

Solution to 4.3.1: The Geometric Change for the two given points is 1.5, so this means that our recursive sequence is:

$$a_n = 1.5 * a_{n-1}, \quad a_0 = 20$$

Example 4.3.2: Given the first two terms of the sequence $a_0 = 100$, $a_1 = 94$, create a recursive sequence for these points using the Arithmetic Change.

Solution to 4.3.2: The Arithmetic Change for the two given points is -6, so this means that our recursive sequence is:

---

[62] As for which one you should use, that depends on the data points. Usually more data is collected before this decision is made.

©Brian Gillispie, 2016.

$$a_n = a_{n-1} - 6, \quad a_0 = 100$$

In both of the previous Examples, we were directly given the two points. Oftentimes this is not the case, and you have to deduce the points from the information given. When that happens, let a_0 represent the first point in time, and let a_1 represent the second point in time. Let's demonstrate this in our next Example.

Example 4.3.3: In the year 2015, your business had $8 million in assets. In the year 2016, your business had $8.4 million in assets. Create a recursive sequence using the Geometric Change that can be used to tell how much in assets you have in a given year.

Solution to 4.3.3: This time, we have to determine which point is the first point and which point is the second point in the sequence. Since the year 2015 comes first, we will let the first point be the assets in the year 2015, which means $a_0 = 8$. Then, this means that the second point is the assets in the year 2016, which means $a_1 = 8.4$.

With our points now defined, we can calculate the desired change. The Geometric Change for these two points is 1.05, which means our desired sequence is:

$$a_n = 1.05 * a_{n-1}, \quad a_0 = 8$$

©Brian Gillispie, 2016.

4.3.2: Sequences from Rates of Change

In practice, it is very rare to only be given two data points to create a sequence from, because with only two points known, it is impossible to know if the sequence should use the Average Rate of Change or the Geometric Change. Instead modelers are either given more points (which makes it more likely that neither will work now), or they are given an initial point and the Rate of Change. When that happens, we can still use the recursive formula as before, but instead plug in our known Rate of Change directly into the formula as appropriate. Let's demonstrate.

Example 4.3.4: Initially, there are 50 students signed up to take English next term, and the number of students signed up for the class is increasing by 4 per week until registration is closed. Create a recursive sequence for this situation.

Solution to 4.3.4: In this problem, we have only an initial value and a Rate of Change. Since our Rate of Change is a rate of increase, this means we were provided with an Average Rate of Change, and that Average Rate of Change is 4. Also, since we only know one point on the sequence, we have no choice but to start at that point. This means $a_0 = 50$ by default, and makes our recursive sequence:

$$a_n = a_{n-1} + 4, \quad a_0 = 50$$

Example 4.3.5: In the year 2019 there are 500 rabbits on a wildlife reserve, and the number of rabbits each year is 250% of the previous number of rabbits. Create a recursive sequence for this situation.

Solution to 4.3.5: This time we are given a percent of, which means we were provided with the Geometric Change. Therefore, our Geometric Change is 2.5 for this problem. Also, we have a starting value, which is the number of rabbits in 2019, so $a_0 = 500$. Putting this into our formula for the recursive sequence we end up with:

$$a_n = 2.5 * a_{n-1}, \quad a_0 = 500$$

©Brian Gillispie, 2016.

Remember that percents can be stated as percent increase or percent decrease too, as we will demonstrate next.

Example 4.3.6: There are currently 100 frogs present, and the number of frogs is increasing by 5% per week. Create a recursive sequence for this situation.

Solution to 4.3.6: The initial number of frogs is 100, which means $a_0 = 100$. Also, we have a percent increase, which means that the Geometric Change is 1.05 by definition of percent increase[63]. Therefore, our desired sequence is:

$$a_n = 1.05 * a_{n-1}, \quad a_0 = 100$$

Be advised that using the Arithmetic or Geometric Change like this only works when there is only one item affecting the next term. If the sequence has both a percent increase and a flat decrease, like say with a payment plan, things are a little more complicated and you will have to reason out how to apply the values. Let's demonstrate.

Example 4.3.7: You bought a $80000 house, and interest is increasing your balance by 0.5% per month, and you are paying $1000 a month on the house. Create a recursive sequence for this situation.

Solution to 4.3.7: This time you have both a percent increase and a flat amount being paid off. Because two items are factoring in, we cannot use the Rate of Changes this time. Instead, we will have to reason out what is going on.

First, note that the house balance is increasing per month by 0.5%. This means that every month, your balance will be 1.005% of the previous months balance. This can be defined recursively as:

$$a_n = 1.005 a_{n-1}$$

---

[63] See Chapter 3 for more on this.

©Brian Gillispie, 2016.

Also, you have a payment of $1000 that is applied to the balance. Payments are always applied after interest is, so your payment of $1000 is subtracted from the new amount of $1.005 * a_{n-1}$. This means that we can then write our final sequence as:

$$a_n = 1.005 * a_{n-1} - 1000, \quad a_n = 80000$$

Notice that while this looks a lot like the Geometric Change and Arithmetic Change were applied at once, that is actually not what is done, as the Geometric Change of that sequence is not 1.005, nor is the Arithmetic Change of that sequence -1000. In this case we had to use another method, one which did not depend on either change. Nevertheless, the ideas we learned on the way through this Section still applied, as we were able to apply the percent increase first, then apply the decrease due to the payment. In general, when given information in the form of Example 4.3.7, percent increases or decreases are always applied first, then the flat increase or decrease is applied second.

©Brian Gillispie, 2016.

4.3 Exercises

1. Given the first two terms of the sequence are $a_0 = 100$, $a_1 = 80$, create a recursive sequence for these points using the Geometric Change.

2. Given the first two terms of the sequence are $a_0 = 76$, $a_1 = 90$, create a recursive sequence for these points using the Geometric Change.

3. Given the first two terms of the sequence are $a_0 = 1000$, $a_1 = 500$, create a recursive sequence for these points using the Geometric Change.

4. Given the first two terms of the sequence are $a_0 = 50$, $a_1 = 55$, create a recursive sequence for these points using the Geometric Change.

5. Repeat Exercise 4, but this time use the Arithmetic Change.

6. Given the first two terms of the sequence are $a_0 = 26$, $a_1 = 30$, create a recursive sequence for these points using the Arithmetic Change.

7. Given the first two terms of the sequence are $a_0 = 40$, $a_1 = 34$, create a recursive sequence for these points using the Arithmetic Change.

8. Given the first two terms of the sequence are $a_0 = 0$, $a_1 = 10$, create a recursive sequence for these points using the Arithmetic Change.

9. In the month of May there are 55 rabbits on an island, and in the month of June there are 66 rabbits on the island. Create a recursive sequence for this situation using the Geometric Change.

10. At 1 pm there were 100 mg of the radioactive substance present, and at 2 pm there were 82 mg of the radioactive substance present. Create a recursive sequence for this situation using the Geometric Change.

©Brian Gillispie, 2016.

11. Currently you have 500 mg of medicine in your body, and the amount present is decreasing by 20% per hour. Create a recursive sequence for this situation.

12. You are making $11 an hour in the year 2016, and your hourly salary is increasing by 5% per year. Create a recursive sequence for this situation.

13. Rent is currently $595 a month in the year 2018, and estimates are your monthly rent will increase by 10% per year. Create a recursive sequence for this situation.

14. As of 5 pm there are 5280 customers in the waterpark, and the number of customers is decreasing by 10 per minute. Create a recursive sequence for this situation.

15. A small town has 250 people as of the year 2016, and the number of people in the town is increasing by 20 per year. Create a recursive sequence for this situation.

16. You have a $100000 house loan that is increasing by 0.0025% per month due to interest, and you are paying $800 per month on the house. Create a recursive sequence for this situation.

17. A river has 14000 salmon, and the number of salmon in the river is increasing by 15% per year. However, harvesting is removing 1000 fish per year from the river as well. Create a recursive sequence for this situation.

18. A small town has 1000 people, and the number of people in the town is increasing by 5% per year. In addition, 100 people per year are moving to the town due to a relocation program. Create a recursive sequence for this situation.

19. For Example 4.3.7, calculate the Arithmetic Change and the Geometric Change between a_0 and a_1. How much do they differ from 1.0025 and -1000?

©Brian Gillispie, 2016.

4.4: Long Term Behavior of Recursive Sequences

So far in this Chapter we have explored how sequences are defined, as well as how to create a recursive sequence from given information. Next, we will begin to look at ways to determine what happens in the long term if this model is unchanged. For instance, if our model is of the spread of a disease, we might wish to know if the disease dies off, or does the disease eventually infect everyone? Knowing this would allow you to determine if some other changes might be called for in order to stop the spread of the disease.

In order to study long term behavior of a model, we first need to study and understand fixed points, which is what we will focus on first in this Section.

4.4.1: Fixed Points of Recursive Sequences

If, in a recursive sequence, the input of the sequence is the same as the output, we say the sequence has a fixed point at that point. For Example, consider the sequence:

$$a_n = 1.02 * a_{n-1} - 200, \quad a_0 = 10000$$

If, we were to compute the value for the term a_1, we would end up with:

$$a_1 = 1.02 * 100000 - 200 = 100000$$

In this case, our value for a_1 is equal to our value for a_0, or, our new value is equal to the current value! Also, it turns out this is not unique to a_1 in that sequence, try to compute a_2, a_3, or even a_{50}, they will all be 10000! In this case, we would say that our sequence has a fixed point at 10000, as the sequence is always at the point 10000.

Next, consider the following sequence:

$$a_n = 1.02 * a_{n-1} - 200, \quad a_0 = 500$$

©Brian Gillispie, 2016.

This is the same sequence with a different starting value. However, now, let's pretend that at some point, one term of this sequence is 100000. Then, if we plug that in to find the new term, we will get 100000 for the new term, the term after that, the term after that, and so on! Therefore, this sequence has the same fixed point as the previous sequence. This also means that the starting value has no impact on what the fixed points are in a sequence, it is entirely dependent on the sequence definition itself.

With all of that stated, we are now ready for the official definition of a fixed point. The official definition of a fixed point is as follows:

> *Definition 4.4.1:* A sequence a has a fixed point r if, once the sequence is on that fixed point it never leaves that point. For a recursively defined sequence, this means that for some finite number n, the following occurs:
>
> $$a_n = a_{n+1} = a_{n+2} = a_{n+3} = a_{n+4} = \cdots$$

For our recursively defined sequences in Sections 4.2 and 4.3, we will be on a fixed point if at any point, $a_n = a_{n-1}$, so we will use this version of the definition going forward.

We will now demonstrate finding fixed points in a few Examples. In none of these examples will we define a starting value, because, as we already saw, the starting value has no bearing on what the fixed point of a sequence is.

Example 4.4.1: Find the fixed points of the sequence $a_n = 0.5 * a_{n-1} + 4$

Solution to 4.4.1: For a fixed point to exist in this sequence, $a_n = a_{n-1} = r$ must occur, where r is the letter we are using to represent our fixed points. Replacing a_n and a_{n-1} by r, since they are both equal to this fixed point once the sequence is on the fixed point, we end up with:

$$r = 0.5 * r + 4$$

Moving all the r terms to one side of the equation gives us:

$$0.5 * r = 4$$

Which, once we divide by 0.5, gives us that:

$$r = 8$$

Therefore, the fixed point to this sequence is 8.

Notice what we did here in the last Example. We found our fixed point by using the definition that a fixed point of a sequence is any point where the input equals the output, then we made that substitution into the equation. By doing this, we lose the subscripts, and can more easily solve the equation for the fixed points.

Example 4.4.2: Find the fixed points of the sequence $a_n = 1.05 * a_{n-1} - 500$

Solution to 4.4.2: By definition of fixed point, the input and output must be equal, or $a_n = a_{n-1} = r$ must occur. Making this substitution for a_n and a_{n-1} results in:

$$r = 1.05 * r - 500$$

Moving both r terms to the same side of the equation makes the equation now:

$$-0.05r = -500$$

Divide both sides by -0.01 to yield the following for the fixed point of the sequence:

$$r = 10{,}000$$

Not every sequence has a fixed point. In fact, most of the problems we worked in Chapter 3 would have a fixed point that isn't 0 if we put them into sequence form. Let's demonstrate that on Example 3.4.3 in sequence form.

Example 4.4.3: Find the fixed points of the sequence $a_n = a_{n-1} + 25$

Solution to 4.4.3: Again, by the definition of the fixed point, the sequence has a fixed point if $a_n = a_{n-1} = r$. Making the substitution for a_n and a_{n-1} gives us:

$$r = r + 25$$

Subtracting r from both sides gives us:

$$0 = 25$$

This is an impossible statement, so we conclude that this sequence has no fixed points, because if it did, then $0 = 25$ would have to be true.

©Brian Gillispie, 2016.

4.4.2: Stability of Fixed Points

Now that we have discussed what fixed points are and how to calculate them, we are ready to look at long term behavior of a sequence. To explore this topic, let's consider the following model which models the spread of a disease, with n in weeks:

$$a_n = 1.05 * a_{n-1} - 500$$

If we solve for the fixed point, we will find that the fixed point of this sequence is 10,000. What this means practically is if, at any point, there are 10,000 people infected with the disease, we will always have 10,000 people infected, as every week we are curing as many as are becoming infected. So the question is, will this sequence end up on that fixed point long term?

In an attempt to see where the sequence ends up long term, we considered two situations, one where we start with 5000 infected, and one where we start with 15000 infected. The results are in a table below, with all numbers rounded to the nearest whole person:

n	$a_0 = 5000$	$a_0 = 15000$
0	5000	15000
1	4750	15250
2	4488	15513
3	4212	15789
4	3923	16079
5	3916	16383

Based on this table, it looks like when we started the sequence with 5000, we moved away from 10000 and are falling towards 0, and when we started the sequence with 15000, we moved away from 10000 and are moving towards infinity. In neither case does the sequence appear to be moving towards the fixed point of 10000. In fact, it appears to be moving away from that fixed point as fast as possible.

Now, consider a different sequence, which is to model the dose of a pain killer in a patient's body in 4 hour intervals, in mg. The sequence is:

$$a_n = 0.5 * a_{n-1} + 500$$

If we solve for the fixed point, we will find that it is 1000. Now, assume you need to take this pain killer for an extended period of time. How much of it will be in your body long term? This time you hope it will not be a really huge number, as that tends to be fatal. Again, we will look at two cases, one where we start with 0 mg of the drug in our body, and one where we started with 1500 mg of the drug already in our body. The results are in the table below:

n	$a_0 = 0$	$a_0 = 1500$
0	0	1500
1	500	1250
2	750	1125
3	875	1062.5
4	937.5	1031.25
5	968.75	1015.625

In the first case, where we started with 0 mg present, it appears that the numbers are increasing, and moving towards the fixed point. In the second case, where we started with 1500 mg present, it appears the numbers are decreasing, and moving towards the fixed point also. In both of these cases, if we plotted the numbers long enough, both scenarios would eventually result in the patient having 1000 mg of the pain killer their body, and that would remain as long as they took the doses of the pain killer every 4 hours.

As you can see, we have two different scenarios that can occur. Our sequence can either converge, or move towards the fixed point, or it can diverge, or move away from, the fixed point. For this reason, we will classify fixed points based on whether the sequence converges or diverges to the fixed points as stable or unstable. The definition of each is as follows:

Definition 4.4.2: A fixed point is **stable** if values that are close to the fixed point converge to the fixed point.

Definition 4.4.3: A fixed point is **unstable** if values that are close to the fixed point do not converge to the fixed point.

How close is close enough to test for stability? The answer is it depends on how many fixed points the sequence has, as well as how chaotic the sequence is. However, it turns out that if the sequence only has one fixed point, any point is sufficient to test for stability. Therefore, for that reason, we will use the starting value of the sequence to test our fixed points for stability, as (almost) all the sequences in this book will have only one fixed point[64]. Just remember though that if your sequence has two or more fixed points, the starting point might not be 'close' enough to the fixed point to give a valid answer though (See [7] and [8] for more on this).

Now, let's demonstrate this technique on a few Examples.

Example 4.4.4: Given the sequence $a_n = 0.5 * a_{n-1} + 4$, $a_0 = 2$, find if the fixed points of the sequence are stable or unstable.

Solution to 4.4.4: First, we need to find the fixed point of the sequence. If we solve for the fixed point[65], we will discover that the fixed point is 8. Now, we need to find if this fixed point is stable or unstable. Since the sequence only has the one fixed point, any point will be sufficient to test the sequence for stability. Therefore, we will plug in the starting value of 2 into the sequence and see what happens. Once we do that, we will compute that:

$$a_1 = 0.5 * 2 + 4 = 5$$

Now, we need to decide if we are moving towards or away from the fixed point. The easiest way to do this is to calculate the distance each point in the sequence is from the fixed point. The distance can be computed by $|a_n - r|$ for each point n, with r the fixed point. Therefore, we first compute the distance from our starting point to r. If we do that, we get that the distance a_0 is from r is $|2 - 8|$ or 6. Therefore, we know we started 6 units away from the fixed point. This means that if a_1 is closer to the fixed point than 6, we are converging on the fixed point, and if a_1 is farther from the fixed point than 6, we are diverging from the fixed point. Plugging a_1 and r into our distance formula gives us $|5 - 8|$, which is 3.

---

[64] Exceptions will be noted as needed.
[65] We solved for this fixed point back in Example 4.4.1.

©Brian Gillispie, 2016.

Since 3 is less than 6, this means the distance from a_1 to the fixed point is less than the distance from a_0 to the fixed point, and therefore the sequence is converging to the fixed point. Therefore, the fixed point of 8 is stable for this sequence.

Example 4.4.5: Given the sequence $a_n = 1.02 * a_{n-1} + 300$, $a_0 = 20000$, find the fixed point of the sequence, and whether the fixed point is stable or unstable.

Solution to 4.4.5: First, we need to find the fixed point. Plug in r for both a_n and a_{n-1} and solve for r, which will result in the following:

$$r = 1.02 * r - 300$$

Or, once we solve that equation, we end up with:

$$r = 15{,}000$$

Next, we need to test if the sequence is converging or diverging to the fixed point. Since we only have one fixed point, we can use any value in the sequence to test the fixed point, so we will test the starting value. Plug in our starting value into the sequence now, and we will discover that the next term is:

$$a_1 = 1.02 * 20000 - 300 = 20{,}100$$

Finally, check the distance a_0 and a_1 are from the fixed point r. If we do that, we will get that the distance from a_0 to r is $|20000 - 15000|$, or 5000, and the distance from a_1 to r is $|20100 - 15000|$, or 5100. Since the distance from the fixed point increased when we went from a_0 to a_1 the sequence is diverging from the fixed point, and the fixed point of 15000 is unstable in this case.

Notice that in the last two Examples, we had a sequence that was increasing from the starting point, but in one case the fixed point was stable, and in the other the fixed point was

©Brian Gillispie, 2016.

unstable. When testing for whether or not a fixed point is stable, do not look at whether the sequence is increasing or decreasing. Instead, look at if the values are getting closer or farther from the fixed point, as that is what matters.

4.4.3: Applications of Stability and Fixed Points

Now that we have seen how to calculate fixed points and how to determine if the points are stable or not, let's use this knowledge to solve a few applications now.

Example 4.4.6: To cure a headache, you decide to take a strong medicine of 500 mg, every 4 hours. Also, every 4 hours, the amount of the medicine in your body decreases by 50%. The medicine is considered to be safe to be in your body until you have 1500 mg present. Is what you are doing safe, or do you need to consider spreading your doses out more?

Solution to 4.4.6: To answer this question, begin by creating the recursive equation to model the situation. Using the rules of Section 4.3, the recursive sequence for this problem is:

$$a_n = 0.5 * a_{n-1} + 500, \quad a_0 = 500$$

Next, find the fixed point. Setting $a_n = a_{n-1} = r$, and making that substitution gives us:

$$r = 0.5 * r + 500$$

Solving for r gives us that our fixed point is:

$$r = 1000$$

Now, we need to figure out what is going on long term. We will plug in our starting value and see if we are moving towards or away from the fixed point. If we plug in our starting dose of 500 into this equation, we will see that one four hour time interval later, we will have:

$$a_1 = 0.5 * 500 + 500 = 750$$

Next, calculate how far both a_0 and a_1 are from the fixed point of 1000. If we do that, we will find that a_0 is $|500 - 1000|$, or 500 away from the fixed point, and a_1 is $|750 - 1000|$, or 250 away from the fixed point. Therefore, we conclude that this sequence is getting closer to the fixed point, and will therefore eventually converge to the fixed point of 1000.

However, we are not done, as we will still need decide if what we are doing is safe or not, based on what we know. Since long term we will end up with 1000 mg of the medicine in our body, and 1000 mg is below the safe limit of 1500 mg, we conclude that what we are doing is safe, and no changes are needed.

Example 4.4.7: You decide to buy a 120,000 house. The bank is loaning you the money at an interest rate which increases the amount of the loan by 0.5% per month, and the loan terms are to pay $595 a month until the house is paid off. Do you ever pay off the house?

Solution to 4.4.7: Start out by setting up the recursive sequence. Using the rules of Section 4.3, our recursive sequence is the following:

$$a_n = 1.005 * a_{n-1} - 595, \quad a_0 = 120000$$

Next, find the fixed points. Letting $r = a_n = a_{n-1}$ and making that substitution gives us:

$$r = 1.005 * r - 595$$

Solve for r find the fixed point of the sequence is:

$$r = 119{,}000$$

Next, test the fixed point for stability. Plug in our starting value into the sequence and we will find that our next value of the sequence is:

4.4: Long Term Behavior of Recursive Sequences

$$a_1 = 1.005 * 120000 - 595 = 120005$$

Next, see how far the starting value and the next value are from the fixed point. The starting value is $|12000 - 119000|$ or 1000 from the fixed point, and the next value is $|120005 - 119000|$, or 1005 from the fixed point. Since the new value is farther from the fixed point than the starting value, as $1005 > 1000$, we conclude that the fixed point is unstable, and that the sequence will move away from the fixed point.

Finally, we need to make our conclusion. Since our sequence is increasing and moving away from the fixed point, we conclude that we will never pay off this house, as the amount of the loan keeps on growing. Also, it probably means it is time to go renegotiate the terms of the loan some.

Example 4.4.8: A rare disease is spreading among the population! Given that there are currently 150,000 infected with the disease, and the disease is growing at the rate of 2% per day, but we are curing 20,000 of the infected people per day, is this disease ever going to be cured at the current rate, or does something else need to be done to ensure that everyone is eventually cured?

Solution to 4.4.8: Start out by creating the model using the rules of Section 4.3. In this case, the disease is growing at the rate of 2% per day, and we are curing 20,000 per day, which gives us:

$$a_n = 1.02 * a_{n-1} - 20000, \quad a_0 = 150{,}000$$

Next, find the fixed points. Making the substitution $r = a_n = a_{n-1}$ in the equation gives us:

$$r = 1.02 * r - 20000$$

Solve this equation for r to find that the fixed point is:

$$r = 1{,}000{,}000$$

Next, test the fixed point for stability. If we plug into our starting value of 150,000, and calculate the next value, we end up with:

$$a_1 = 1.02 * 150{,}000 - 20{,}000 = 133{,}000$$

Next, check the distances of the known points. The starting value is 850,000 away from the fixed point, and the next value is 867,000 away from the fixed point, which means the distance from the fixed point is growing. This means then that the fixed point of 1,000,000 is unstable, and the sequence will be pushed away from that point.

Finally, we need to make a conclusion, based on what we know. Since the sequence is moving away from the fixed point of 1,000,000, and the sequence is also decreasing, this means that at some point, the sequence will have a value of 0. This means that at some point in the future, everyone will be cured of the disease.

4.4 Exercises

For the sequences in Exercises 1 - 12, find the fixed points, then classify the fixed points as stable or unstable. If no fixed points exist, state so.

1. $a_n = 0.8 * a_{n-1} + 600, \quad a_0 = 1000$

2. $a_n = 1.003 * a_{n-1} - 750, \quad a_0 = 50000$

3. $a_n = 1.2 * a_{n-1} - 1000, \quad a_0 = 7500$

4. $a_n = 0.78 * a_{n-1} + 650, \quad a_0 = 3500$

5. $a_n = 1.002 * a_{n-1} - 500, \quad a_0 = 60000$

6. $a_n = 0.6 * a_{n-1} + 1000, \quad a_0 = 1250$

7. $a_n = 0.6 * a_{n-1} + 1000, \quad a_0 = 5000$

8. $a_n = a_{n-1} + 10$

9. $a_n = 0.85 * a_{n-1} + 750, \quad a_0 = 7500$

10. $a_n = 1.1 * a_{n-1} - 5000, \quad a_n = 120000$

11. $a_n = 1.15 * a_{n-1} + 250, \quad a_0 = 50$

12. $a_n = 0.82 * a_{n-1} + 500, \quad a_0 = 0$

Using the techniques of this (and previous) Sections, answer the following question. Please state your recursive sequence and explain how you arrived at your answer.

13. A doctor wishes to give a patient a dose of 500 mg a day indefinitely. If the drug decreases by 71% per day, find the amount of drug present in the patient long term.

14. You win a large sum of money, and decide to invest $500,000 of it into an account that will increase your money by 1% per month. If you also decide to withdraw $7500 per month, in the long term how much money will you have in your account?

15. A new disease is killing off the rabbit population in the area! There are estimated 120,000,000 rabbits in the area, and the rabbits are decreasing by 5% per day. However, 100 new rabbis are immigrating into the area per day too. In the long term, do the rabbits go extinct?

16. A pollutant is arriving into a lake at the rate of 0.5 liter per day. If the filtration techniques on the lake are able to remove 95% of the pollutant per day, and as long as the pollutant stays below 1 liter it is still safe to swim in the lake, in the long run, will the lake be safe to swim in?

17. A herd of wolves have arrived in the area. Wolves are growing at the rate of 3% per year (after deaths are accounted for), and they also gain 5 wolves a year due to new packs showing up. If the pack originally has 25 wolves, long term how many wolves will there be?

18. A model for a new disease outbreak concludes that we are curing the disease at the rate of 5% of the infected people per week, but every week 1000 new people are infected. Currently there are 500,000 infected, and if the number of infected ever drop below 7,500 the disease will die off naturally. Does the disease ever die off at this rate?

©Brian Gillispie, 2016.

19. You have attempted to reintroduce dragons to the far northern wilds of Canada! You introduce 50 dragons, but every year 2% of the dragons die off. Using whatever method allowed you to reintroduce dragons, you can create another 4 dragons every year. For the dragons to survive on their own, it is estimated that there needs to be 100 dragons present at one point in time. Does this ever happen?

20. You currently manage a burger joint. Every day, your sales increase by 3%, but also every day 12 people give up eating burgers for various reasons. If you currently sell 500 burgers a day, long term how many burgers will you sell?

21. A town has a population of 50,000 people. Every year the population increases by 2% due to natural causes, but also every year 250 people move out of the town. The town has been promised a new super highway if the population ever exceeds 75,000. Does this town ever get the new highway?

22. Zombies have attacked! When the zombies attack your town there are 100,000,000 zombies attacking, and they are growing at the rate of 2% per hour. However, you are managing to kill off 25,000 zombies per hour. Do you ever manage to kill off all the zombies, or would you be better off running away as fast as you can?

23. A disease outbreak has occurred! Given that there are currently 1,000,000 people infected with the disease, and every week the number of infected increases by 2%, if we are currently curing 50,000 of the disease every week, long term, how many will be infected by this disease?

24. You are monitoring the wolf population on an island. At the moment there are 2,500 wolves, and the population is increasing at the rate of 3% per year. However, every year, 125 wolves are hunted and killed by rouge hunters. If the number of wolves on the island ever drops below 150 wolves, wolves will become extinct due to being unable to find mates easily. If nothing changes, do the wolves eventually go extinct on this island?

©Brian Gillispie, 2016.

4.5: Interconnected Sequences

So far in this Chapter we have discussed sequences, and used our sequenced to make various predictions. However, our sequences so far have assumed that all the outside factors for our sequence are constant over time. For example, when we modeled the population of rabbits, we assumed that the growth rate of the rabbits is the same no matter if it is month one or month one hundred of the sequence. In reality, the growth rate of the rabbits might be affected by the number of predators in the area, which are modeled by their own sequence as well!

When this occurs, we have what are called interconnected sequences, which are sequences of the following form, with k_1 through k_6 all constants:

$$a_n = k_1 * a_{n-1} + k_2 * b_{n-1} + k_3$$
$$b_n = k_4 * a_{n-1} + k_5 * b_{n-1} + k_6$$

Notice how the sequence for A depends on both the previous value for A and the previous value for B? That is what is meant by an interconnected sequence. Both sequences depend on the other's previous values, and therefore you will have to compute both the values of the sequence A and the sequence B as you proceed. Let's demonstrate.

Example 4.5.1: Given the following predator-prey equations[66], with n in years:

$$r_n = 1.02 * r_{n-1} - 0.001 * r_{n-1} * w_{n-1}$$
$$w_n = 0.9 * w_{n-1} + 0.001 * r_{n-1} * w_{n-1}$$

If we currently have 75 rabbits and 15 wolves, how many of each will we have after one year passes?

---

[66] These are called predator-prey because one of the animals in one of the sequences eats the other.

©Brian Gillispie, 2016.

Solution to 4.5.1: Since the sequence is asking us how many wolves and rabbits are present after one year passes, we need to determine the value of r_1 and w_1 respectively. By the definition of the sequences, we have the following for r_1 and w_1:

$$r_1 = 1.02 * r_0 - 0.001 * r_0 * w_0$$
$$w_1 = 0.9 * w_0 + 0.001 * r_0 * w_0$$

Substituting our given starting number of wolves and rabbits in for r_0 and w_0 gives us:

$$r_1 = 1.02 * 75 - 0.001 * 75 * 15$$
$$w_1 = 0.9 * 15 + 0.001 * 75 * 15$$

Finish the calculations to get the following:

$$r_1 = 75.375$$
$$w_1 = 14.625$$

Which means, after one year passes, we will have about[67] 75 rabbits and 15 wolves present.

Example 4.5.2: Given the following predator-prey equations for the number of owls and mice in the area, with n in years:

$$o_n = 0.975 * o_{n-1} + 0.00001 * o_{n-1} * m_{n-1}$$
$$m_n = 1.4 * m_{n-1} - 0.001 o_{n-1} * m_{n-1}$$

If there are currently 1500 mice and 50 owls, how many mice and owls will be present after one year passes?

---

[67] We say about as we have to interpret our decimal answer somehow.

©Brian Gillispie, 2016.

Solution to 4.5.2: Just like before, we wish to know how many of each species are present after one year passes, of, the value of o_1 and m_1 in these sequences. Plugging in our starting values for o_0 and m_0 gives us:

$$o_1 = 0.975 * 50 + 0.00001 * 50 * 1500$$
$$m_1 = 1.4 * 1500 - 0.001 * 50 * 1500$$

Which, computes to:

$$o_1 = 49.5$$
$$m_1 = 2092.5$$

Which means that after one year has passed, there will be about 49 to 50 owls present, and about 2092 to 2093 mice present.

Example 4.5.3: Using the same sequence in Example 4.5.2, how many mice and owls will be present after 2 years pass?

Solution to 4.5.3: This time, we are being asked to compute the value of o_2 and m_2, which depend on the values of o_1 and m_1. We already computed those in the last Example, so we will use those values here[68]. Plugging them into our sequences gives us:

$$o_1 = 0.975 * 49.5 + 0.00001 * 49.5 * 2092.5$$
$$m_1 = 1.4 * 2092.5 - 0.001 * 49.5 * 2092.5$$

Which, computes to:

$$o_2 = 49.2982875$$

---

[68] It is worth pointing out that we will be using the values of o_1 and m_1 before rounding. This is to avoid introducing roundoff error. Try redoing this exercise and rounding your answers to Example 4.5.2 to 49 and 2093 and see how it affects your answer to Example 4.5.3.

©Brian Gillispie, 2016.

$$m_2 = 2825.92125$$

Or, after two years have passed, there are about 49 owls and 2826 mice present in the area.

Example 4.5.4: Given the following disease model, with sequence S representing the number who are susceptible to the disease, and sequence I representing those who are infected with the disease per week, with n in weeks:

$$s_n = 0.95 * s_{n-1} + 0.15 * i_{n-1}$$
$$i_n = 0.85 * i_{n-1} + 0.05 s_{n-1}$$

If we currently have 100,000 infected and 900,000 who are susceptible, how many infected and susceptible will we have after two weeks pass?

Solution to 4.5.4: Start off by plugging in our starting values into both sequences. If we do that, we end up with:

$$s_1 = 0.95 * 900000 + 0.15 * 100000$$
$$i_1 = 0.85 * 100000 + 0.05 * 900000$$

Which, calculates to:

$$s_1 = 870000$$
$$i_1 = 130000$$

Now that we know the values of s_1 and i_1, we can find our desired values of s_2 and i_2. Plugging into the sequence again gives us:

$$s_2 = 0.95 * 870000 + 0.15 * 130000$$
$$i_2 = 0.85 * 130000 + 0.05 * 870000$$

©Brian Gillispie, 2016.

Which, calculates to:

$$s_2 = 846000$$
$$i_2 = 154000$$

Which means that, after 2 weeks have passed, there are 154,000 infected and 846,000 susceptible to the disease.

Interconnected sequences can have three or even more sequences, all connected together. A common such interconnected sequence is the S-I-R Model, which we shall use in the next Example.

Example 4.5.5: To model the spread of a disease, the S-I-R model is used, with the three sequences S for the number of susceptible, I for the number of infected, and R for the number of released[69]. The three sequences are as follows, with n in weeks:

$$s_n = 0.95 * s_{n-1} + 0.10 * i_{n-1}$$
$$i_n = 0.75 * i_{n-1} + 0.05 * s_{n-1}$$
$$r_n = 0.15 * i_{n-1} + r_{n-1}$$

If we currently have 750,000 susceptible, 50,000 infected, and none that are released, how many susceptible, infected, and released will we have after 2 weeks pass?

Solution to 4.5.5: Even though we have three sequences this time, the steps are the same as for when we had two sequences. Begin by plugging the starting values for each into the sequences to compute how many of each are present after one week has passed, then use that number to calculate how many are present after two weeks have passed. Plugging in our starting values into the sequences gives us:

---

[69] Released means they have left the model. This usually occurs either by death or by gaining immunity to the disease.

©Brian Gillispie, 2016.

$$s_1 = 0.95 * 750000 + 0.10 * 50000$$
$$i_1 = 0.75 * 50000 + 0.05 * 750000$$
$$r_1 = 0.15 * 50000 + 0$$

Which, calculates to:

$$s_1 = 717500$$
$$i_1 = 75000$$
$$r_1 = 7500$$

Next, plug these values into the sequence to find how many of each group there are on week two. If we do that, we end up with:

$$s_2 = 0.95 * 717500 + 0.10 * 75000$$
$$i_2 = 0.75 * 75000 + 0.05 * 717500$$
$$r_2 = 0.15 * 75000 + 7500$$

Which, calculates to:

$$s_2 = 689125$$
$$i_2 = 92125$$
$$r_2 = 18750$$

Therefore, we can conclude that in two weeks we will have 689,125 susceptible, 92,125 infected, and 18,750 released, based on this model.

While it is possible to determine the long term behavior of interconnected sequences, doing so requires techniques from linear algebra except in very special cases (See [8]), and for that reason we will not study long term behavior of these sequences at this point in time.

©Brian Gillispie, 2016.

4.5 Exercises

1. For our model in Example 4.5.1, compute how many rabbits and wolves will be present in year two. Do not round the answers from year one.

2. For our model in Example 4.5.1, compute how many rabbits and wolves will be present in year two, rounding your answers from year one to 15 wolves and 75 rabbits.

3. If the answer in Exercise 1 is the actual answer, and the answer in Exercise 2 is the calculated answer, what is the relative error of your answer for the number of rabbits present?

4. For our model in Example 4.5.1, compute how many rabbits and wolves will be present in year five. *Hint: You may wish to calculate these in Excel*

5. Given the following equation for the number of wolves and rabbits in an ecosystem, with n in years:

$$r_n = 1.2 * r_{n-1} - 0.0025 * r_{n-1} * w_{n-1}$$
$$w_n = 0.97 * w_{n-1} + 0.000001 * r_{n-1} * w_{n-1}$$

If there are currently 75 wolves and 2500 rabbits, how many of each species will be present after one year passes?

6. Using the system in Exercise 5, calculate how many of each will be present after two years pass.

7. Using the system in Exercise 5, if we instead had 10 wolves and 60,000 rabbits present, calculate how many of each will be present after one year passes?

8. Using the system in Exercise 5, if we instead had 100 wolves and 20,000 rabbits present, calculate how many of each will be present after two years pass?

©Brian Gillispie, 2016.

9. Using the system in Exercise 5, if we instead had 500 wolves and 1,000,000 rabbits present, calculate how many of each will be present after one year passes.

10. Sometimes, two species can assist each other in surviving an ecosystem. Pretend we have two bacteria, which we will call bacteria A and B. The number of cells of each bacteria present after n hours have passed are given as follows, with n in hours:

$$a_n = 1.1 * a_{n-1} + 0.00001 * a_{n-1} * b_{n-1}$$
$$b_n = 1.05 * b_{n-1} + 0.002 * a_{n-1} * b_{n-1}$$

If we initially have 5 cells of bacteria A and 10 cells of bacteria B, how many of each will be present after one hour passes?

11. For the system in Exercise 10, how many of each cell will be present after six hours pass?

12. For the system in Exercise 10, if we instead had 200 cells of bacteria A and 50 cells of bacteria B, how many of each will be present after one hour passes?

13. Two competing stores wish to know how many customers will come to their store each week, and have devised the following sequences to predict the number of customers, with S and C representing the names of the stores, and with n in weeks.

$$S_n = 0.80 * S_{n-1} + 0.15 * C_{n-1}$$
$$C_n = 0.20 * S_{n-1} + 0.85 * C_{n-1}$$

If store S has 1000 customers and store C has 800 customers, how many customers will each store have one week from now?

14. Repeat Exercise 13, but this time compute how many customers each store will have three weeks from now.

©Brian Gillispie, 2016.

15. Repeat Exercise 13, but this time compute how many customers each store will have 20 weeks from now. *We highly suggest using excel for this problem!*

16. Repeat Exercise 13, but this time Store S has 0 customers and Store T has 1800 customers. Compute how many each store will have 20 weeks from now.

17. A highly infectious flu virus has broken out in a town. The following interconnected sequences were created via an S-I-R model to model the number of susceptible, infected and released that were present, with n in weeks:

$$s_n = 0.5 * s_{n-1} + 0.02 * i_{n-1}$$
$$i_n = 0.46 * s_{n-1} + 0.3 * i_{n-1}$$
$$r_n = 0.04 * s_{n-1} + 0.68 * i_{n-1} + r_{n-1}$$

If there are currently 15,000 people in the town, and all of them are susceptible, how many people are susceptible, infected, and released one week later?

18. Repeat Exercise 17, but this time compute how many are susceptible, infected, and released five weeks later.

19. Deduce how many susceptible, infected, and released there will be long term.

©Brian Gillispie, 2016.

20. The following is an S-I-R model for the cold virus during cold and flu season, with n in weeks.

$$s_n = 0.85 * s_{n-1} + 0.02 * i_{n-1} + 0.2 * r_{n-1}$$
$$i_n = 0.10 * s_{n-1} + 0.01 * i_{n-1}$$
$$r_n = 0.05 * s_{n-1} + 0.97 * i_{n-1} + 0.8 * r_{n-1}$$

If initially there are 6,000 people, all of them susceptible, how many will be susceptible, infected, and released two weeks later?

21. Repeat Exercise 20, but this time, calculate how many will be susceptible, infected, and released four weeks later.

22. Using excel or some other method, and using the model in Exercise 20, calculate how many are susceptible, infected, and released 26 weeks later at the end of the cold season.

23. Repeat Exercise 20, but this time there are 5800 susceptible and 200 infected initially, and calculate how many are susceptible, infected and released two weeks later.

24. A deadly outbreak has occurred! The following is a model for how many are susceptible, infected, and released for this disease, with time in weeks.

$$s_n = 0.7 * s_{n-1} + 0.02 * i_{n-1}$$
$$i_n = 0.28 * s_{n-1} + 0.48 * i_{n-1}$$
$$r_n = 0.02 * s_{n-1} + 0.50 * i_{n-1} + r_{n-1}$$

If there are currently 299,999,990 susceptible, 10 infected, and no released, calculate how many are susceptible, infected and released one week from now.

©Brian Gillispie, 2016.

Chapter 4 Summary

4.1: In this Section, we defined what a sequence was, and listed the common terms used with sequences. A sequence is a list of ordered objects, often referenced with a lower case letter. a_n stands for the nth element in the ordered sequence. Sequences can be denoted by either spelling out all of the terms in the sequence, or by explicit formulas.

4.2: In this Section, we discussed what recursive sequences are. Recursive sequences are sequences where the current term of the sequence depends on one or more of the previous values. We often denote the starting term of the sequence as a_0, due to the fact that the starting term usually is referencing the value of the sequence at time zero.

4.3: In this Section, we discussed how to make recursive sequences from the given data. If two points are given, you can use either the Arithmetic or the Geometric Change to create a sequence, and if only an initial value is given, then you have to use the Rate of Change that is known to create the appropriate sequence.

4.4: In this Section we discussed fixed points of sequences. A fixed point of a sequence is the point where the inputs and outputs of the sequence are equal. If a sequence is affine, or of the form $a_n = c * a_{n-1} + d$, then the fixed point r is the point where $a_n = a_{n-1} = r$. Fixed points are independent of the starting value, so no starting value needs to be provided to find one.

Fixed points can either be stable or unstable. A fixed point is stable if values that start close to the fixed point move towards the fixed point. A fixed point is unstable if values that start close to the fixed point move away from the fixed point.

4.5: In this Section we discussed interconnected sequences, or sequences where the next term of the sequence depends on the previous term of another sequence. To calculate the values of interconnected sequences, you need to compute the values of each step of the sequence in turn.

©Brian Gillispie, 2016.

Chapter 5: Linear Models

In this Chapter we will discuss how to create and use our first continuous model, the linear model. We will start off by reviewing the common terms associated with lines, and how we can create a line with two known data points. From there, we will discuss how to use a linear model with our given information. Then, we will close with a couple Examples of real world applications of linear models.

©Brian Gillispie, 2016.

©Brian Gillispie, 2016.

5.1: Review of Lines

Before we can begin working with linear models, we need to review how to create the equation of a line. In this Section we will discuss common terms associated with lines, and from there we will proceed to discuss the various equations of a line. The reader who is familiar with these concepts can skip ahead to Section 5.2, and return here as needed.

5.1.1: Slope

The first term we will define in this Section is the slope of a line, which is defined as follows:

Definition 5.1.1: If the points (x_1, y_1) and (x_2, y_2) are on the same line, then the slope of that line is defined as follows:

$$m = \frac{\text{rate of change of the dependent variable}}{\text{rate of change of the independent variable}}$$

This is more commonly written as:

$$m = \frac{y_2 - y_1}{x_2 - x_1}$$

This slope definition is more commonly known as the change in y divided by the change in x, or the rise over run. Also, note the subscripts by each letter. The subscript by the letter tells you which point the data value comes from. So, for instance, x_2 means the x value from the second point. Which point is point one and point two is up to the modeler, but in this book we will use the first point listed in the problem (assuming reading everything in order), as our first point, and the next point as the second point, and so on, unless noted otherwise.

©Brian Gillispie, 2016.

With the slope of a line now defined, let's now demonstrate with a few Examples how to calculate the slope of a line connecting two points:

Example 5.1.1: Calculate the slope of a line containing the points $(2, 4)$ and $(7, 14)$.

Solution to 5.1.1: In this problem, $(2, 4)$ is the first point, and $(7, 14)$ is the second point. Plugging those into the slope equation gives us the following:

$$m = \frac{14 - 4}{7 - 2} = \frac{10}{5} = 2$$

Therefore, the slope of the line containing these points is 2.

Be careful when plugging in your points into the slope equation, especially if the points are negative. A common error is to forget to include the minus sign from the point itself in the equation. However, the equation does require you to subtract whatever that point is, even if the point is negative. Let's demonstrate now with an Example where a couple of the coordinates are negative.

Example 5.1.2: Calculate the slope of a line containing the points $(-5, 15)$ and $(2, -6)$.

Solution to 5.1.2: In this problem, $(-5, 15)$ is the first point, and $(2, -6)$ is the second point. Plugging those into the slope equation, and being careful with our minus signs, gives us the following:

$$m = \frac{-6 - (15)}{2 - (-5)} = \frac{-21}{7} = -3$$

Therefore, the slope of the line containing these points is -3.

Notice we used () when plugging in the first point into the slope equation. This allowed us to better see that in the denominator that we had $2 - (-5)$ for our equation. A common

©Brian Gillispie, 2016.

mistake is to accidentally plug in $2 - 5$ in the denominator, which is now saying that your first point is $(5, 15)$, not $(-5, 15)$ as stated. Always include the minus sign from the points.

Another thing that can arise when computing the slope is after the initial calculations you end up with a zero in your fraction. Let's demonstrate now a couple of Examples where that happens.

Example 5.1.3: Calculate the slope of the line containing the points $(0, 10)$ and $(1, 10)$.

Solution to 5.1.3: In this problem, $(0, 10)$ is the first point, and $(1, 10)$ is the second point. Plugging those into the slope equation gives us the following:

$$m = \frac{10 - 10}{1 - 0} = \frac{0}{1} = 0$$

Therefore, the slope of the the line containing these two points is 0.

Example 5.1.4: Calculate the slope of the line containing the points $(5, 25)$ and $(5, 30)$.

Solution to 5.1.4: In this problem, $(5, 25)$ is the first point and $(5, 30)$ is the second point. Plugging those into the slope equation gives us the following:

$$m = \frac{30 - 25}{5 - 5} = \frac{5}{0}$$

However, we cannot divide by 0, so we say that the slope of this line is undefined.

As the last four Examples have demonstrated, the slope of a line can be either positive, negative, zero, or undefined. This also means that no slope is an invalid answer. All sets of two points connected by a line have a slope, unless the slope is undefined.

©Brian Gillispie, 2016.

5.1.2: The Y Intercept

Now that we have discussed how to compute the slope of a line, we will turn our attention to the second definition used when working with lines, which is the y intercept[70].

Definition 5.1.2: The **y intercept** of a line is the y value of the equation when x is equal to 0.

Let's demonstrate this definition with an Example.

Example 5.1.5: Given the equation $y = 4x + 50$, what is the y intercept for this equation?

Solution to 5.1.5: The y intercept is defined as the value of y when x is equal to zero. Therefore, to find the y intercept, plug $x = 0$ into the equation. Doing this gives us:

$$y = 50$$

Therefore, the y intercept of this equation is 50.

By Definition 5.1.2, the y intercept is defined for more than just lines. In fact, any equation where $x = 0$ is in the domain has an y intercept! We'll demonstrate that in our next Example.

Example 5.1.6: Given the equation $y = -16x^2 + 64x + 5$, what is the y intercept for this equation?

Solution to 5.1.6: Since the y intercept is defined as the value of y when x is equal to zero, we will plug in $x = 0$ into the equation. Doing this gives us:

$$y = -16 * 0^2 + 64(0) + 5 = 5$$

---

[70] Technically the y intercept is a point, but we usually only refer to the y coordinate of the point as the y intercept.

©Brian Gillispie, 2016.

Therefore, the *y* intercept of this equation is 5.

5.1.3: Definition of a Line

With both slope and *y* intercept defined, we are now ready to define the equation of a line. The standard equation of a line is as follows[71]:

> *Definition 5.1.3:* A **line** is an equation of the form $y = mx + b$, where *m* is the slope of the line, and *b* is the *y* coordinate of the *y* intercept of the line

Based on these definitions, in order to create an equation of a line, we need to be able to calculate the slope of the line, and we need to know what the *y* intercept is. Let's now demonstrate this with a few examples:

Example 5.1.7: Find the equation of the line containing the two points $(3, 15)$ and $(0, 9)$.

Solution to 5.1.7: Using $(3, 15)$ as the first point, and $(0, 9)$ as the second point, and plugging those into the slope equation, we find the following for the slope:

$$m = \frac{9 - 15}{0 - 3} = \frac{-6}{-3} = 2$$

Now, we need to figure out what the *y* intercept is. We are given the point $(0, 9)$ which says that when *x* is 0, *y* is 9. The *y* intercept is defined as the *y* coordinate when $x = 0$, so that means that our *y* intercept is 9.

---

[71] Some books use $Ax + By = C$ as the standard. However, that form is pretty useless for us when modeling as we would always have to solve it for *x* or *y* to use it for anything. For that reason, we will not be using that equation of a line in this book.

©Brian Gillispie, 2016.

To finish the problem, we need to plug everything into our equation for a line. The slope is 2, and the y intercept is 9, so we let b be 9 and m be 2 in the formula, which gives us the following for the final equation of the line:

$$y = 2x + 9$$

In the last example, our y intercept was one of the given points. Oftentimes, that is not the case. When that happens, you will have to calculate the y intercept in order to use this equation of a line. Let's demonstrate.

Example 5.1.8: Find the equation of the line containing the two points $(-1, 10)$ and $(2, 25)$.

Solution to 5.1.8: Using $(-1, 10)$ as the first point and $(2, 25)$ as the second point, and plugging those into the slope equation, we find the following for the slope:

$$m = \frac{25 - 10}{2 - (-1)} = \frac{15}{3} = 5$$

Next, we need to figure out what the y intercept is, but the y intercept is not given this time. To find the y intercept, begin by plugging the slope into the equation for a line. Once that is done, we end up with:

$$y = 5x + b$$

Now, since the equation of a line has to be defined for all points, including the two we were given, we can plug into the equation of the line one of the two given points (which one does not matter, just pick one). Plugging in the point $(2, 25)$ into our equation, we end up with:

$$25 = 5(2) + b$$

This is an equation we can solve. Solve this equation for b to end up with:

©Brian Gillispie, 2016.

$$b = 15$$

To finish, we need to state our final equation of a line. Our slope was 5, and our y intercept was 15. Plugging those into the equation of a line gives us the following for our final equation of a line:

$$y = 5x + 15$$

5.1.4: Point-Slope Form of a Line

As the reader can see, if we use Definition 5.1.3 for our equation of a line, and if the y intercept is not given to the modeler, some extra work is often required to find the y intercept. However, there is another way to create an equation of a line that does not require all that extra work, and that is we could use the point-slope equation of the line instead. The definition of point-slope form of a line is as follows:

> *Definition 5.1.4:* The **point-slope equation of a line** is an equation of the following form:
>
> $$y - y_1 = m(x - x_1)$$
>
> Where m is the slope of the line, and (x_1, y_1) is any point on the line.

The point-slope equation of a line only requires that the modeler know the slope and one of the points on the line. This makes the point slope form of a line easier to work with when the y intercept is not in the given data. The trade-off is that point-slope form is not as easy to work with when trying to make predictions, but that can be fixed by solving the equation for y first.

Let's now create a couple of equations in point-slope form to demonstrate:

Example 5.1.9: Find the equation of a line containing the two points $(3, 30)$ and $(4, 45)$.

Solution to 5.1.9: Using $(3, 30)$ as the first point and $(4, 45)$ as the second point, and plugging those into the slope equation gives us the following for our slope:

$$m = \frac{45 - 30}{4 - 3} = 15$$

Plugging this slope into the point-slope equation, and using the point $(3, 30)$ gives us the following:

$$y - 30 = 15(x - 3)$$

As the reader can see, creating the equation in point-slope form is much faster when the y intercept is not given. It does have the disadvantage of being in a slightly less familiar form, but if one wishes to, one can do the algebra and solve the equation for y to change the equation from point slope form into slope intercept form.

Example 5.1.10: Find the equation of the line containing the two points $(80, 1300)$ and $(100, 1500)$.

Solution to 5.1.10: Using $(80, 1300)$ as the first point, and $(100, 1500)$ as the second point, and plugging those points into the slope equation gives us the following for our slope:

$$m = \frac{1500 - 1300}{100 - 80} = \frac{200}{20} = 10$$

Plugging the slope into our point-slope equation, and using the point $(100, 1500)$ gives us the following:

$$y - 1500 = 10(x - 100)$$

©Brian Gillispie, 2016.

Notice that this time I used point two in the point-slope equation instead of point one. You are not required to use point one in the point slope equation if you don't wish to, as the equation for point-slope form only requires that you plug in a point that is on the line. So which point you plug in into the point slope equation is up to you.

However, because the notation of the point slope equation looks a lot like the slope equation notation, many choose to use the first point in the point-slope form to avoid confusion. Which you do is up to you.

©Brian Gillispie, 2016.

5.1 Exercises

For Exercises 1 - 4, find the y intercept of the following equation

1. $y = -25x + 100$

2. $y = 7x + 92$

3. $y = -16x^2 + 80x + 25$

4. $y = 10 * 2^x$

Find the equation of the line connecting the following points:

5. $(2, 5)$ and $(4, 65)$

6. $(1, 1000)$ and $(6, 11000)$

7. $(0, 25)$ and $(2, 13)$

8. $(2, 1000)$ and $(5, 350)$

9. $(2000, 5.6)$ and $(2005, 4.3)$

10. $(212, 0)$ and $(199, 10000)$

11. $(1, 50)$ and $(2, 50)$

12. $(0, 15)$ and $(0, 30)$

©Brian Gillispie, 2016.

Create an equation of a line for the following problems. Clearly define what x and y represents in your equations.

13. The speed of sound is 337 meters per second at 10 degrees Celsius and 343 meters per second at 20 degrees Celsius. Create an equation that can predict the speed of sound based on the current temperature in Celsius.

14. It is 0 degrees Celsius when it is 32 degrees Fahrenheit, and it is 100 degrees Celsius when it is 212 degrees Fahrenheit. Create an equation that will tell you the Fahrenheit temperature based on the current Celsius temperature.

15. It costs a printing company $250 dollars to print 10 books, and $350 dollars to print 20 books. Create an equation that will tell you how much it costs to print, based on the number of books you wish to print.

16. A wildlife reserve finds that they have 100 ducks in the year 2015, and 120 ducks in the year 2017. Create an equation that will predict how many ducks they will have in a given year.

17. An island has 56 goats in the year 2020, and 125 goats in the year 2030. Create an equation that will tell them how many goats they will have in a given year.

18. When gas is $2.89 a gallon, 15000 gallons of gas are sold, and when gas is $2.59 a gallon, 21,000 gallons of gas is sold. Create an equation that will predict how many gallons of gas are sold based on the price per gallon.

©Brian Gillispie, 2016.

5.2: Modeling with Slopes and Intercepts

In the previous Section we discussed the terms used when creating a linear model. In this Section we will now discuss how we can create a linear model if we know the rate of change as well as the starting value for the situation being modeled. Before we can do that though, we need to better understand what the various terms mean in the context of our equation.

5.2.1: Interpreting Slope

To understand the slope of a line, let's start out by looking at the slope equation. Recall that the slope equation for the slope of a line through two points $(x_1, y_1), (x_2, y_2)$ is:

$$m = \frac{y_2 - y_1}{x_2 - x_1}$$

Now, recall also that the formula for the Average Rate of Change between two points $(x_1, y_1), (x_2, y_2)$ is:

$$ARC = \frac{y_2 - y_1}{x_2 - x_1}$$

Notice that they are the exact same formula? That means that the slope of a line can also be thought of as how much y is changing every time we change x by one unit. Therefore, any time we know that something is increasing by so much per day, or decreasing by so much per week, that number is our slope. Let's demonstrate with a few Examples.

©Brian Gillispie, 2016.

Example 5.2.1: A lake is currently at 30 feet, and the water level is rising by 0.15 feet per day. What is the slope of the linear equation that represents this situation?

Solution to 5.2.1: In this case, notice that the lake level is changing by 0.15 feet per day. Therefore, that number is our slope, and the slope in our linear model will be:

$$m = 0.15 \text{ feet/day}$$

Example 5.2.2: In October there are 126 ducks on a pond, but due to migration the number of ducks on the pond is decreasing by 50 ducks per month. What is the slope of the linear equation that represents this situation?

Solution to 5.2.2: In this problem, the number of ducks is decreasing by 50 ducks per month. Therefore, that number is our slope, and the slope in our linear model will be:

$$m = -50 \text{ ducks/month}$$

Notice the slope is negative because the number of ducks is decreasing every month.

5.2.2: Interpreting Y - Intercepts

Next we need to consider what the y intercept means in the context of a linear model. By definition, the y intercept of a linear equation is the y value when x is zero. However, since x is usually time when modeling, then a x value of zero corresponds to the time the model started, which means that the y value at that point is the starting amount. Therefore, what this means is if we know the starting value, we can use that as our y intercept! Let's demonstrate.

Example 5.2.3: Initially, there are 30 ducks on a pond, and the number of ducks is decreasing by 1 per week. What is the y intercept for a linear model for this situation?

Solution to 5.2.3: We want the y intercept to be whatever the starting value is for the model. Since there were initially 30 ducks, we should use 30 for our y intercept then.

Example 5.2.4: 50 goats were released into the wild. Each year the number of goats is increasing by 2 goats per year. What is the y intercept for a linear model for this situation?

Solution to 5.2.4: In this problem, we know that 50 goats were released into the wild, which means we start out with 50 goats. Therefore, we should use 50 for our y intercept.

Notice what we did here. In both examples, we looked for something that told us what the starting amount was, and made that value our y intercept. This will always work provided the data given provides you with an initial or starting amount.

5.2.3: Putting it all Together

Now that we have seen how to find the slope and y intercept from a given problem, let's work a few Examples where we find and use a linear model.

Example 5.2.5: Initially there were 50 wolves released into the wild, and the number of wolves is increasing by 2 wolves per month. How many wolves will be present after 3 months pass?

Solution to 5.2.5: First we need to figure out what is our slope and y intercept for this problem. The slope is the rate of change, which is 2 wolves per month, and the y intercept is the initial number of wolves, which is 50. Therefore, that means our equation is:

$$y = 2x + 50$$

Next, we need to answer the question, which is how many wolves will be present when 3 months pass. However, to do that, we need to know what x and y represent in the problem! The easiest way to know what x and y represent in a linear model is y is always in the same units as the y intercept[72], and x is in whatever units are needed so that the addition step is legal in the problem. Since our slope is in the units wolves per month, then x must be whatever units are required so that:

$$\frac{wolves}{month} * units\ of\ x = wolves$$

The only way for that to happen is x is in months[73]. Therefore, since we know x is in months, plug in 3 for x and we will end up with the answer of:

$$y = 2(3) + 50 = 56$$

Which means after 3 months pass, 56 wolves are present.

As you can see, the toughest part of this is determining the units of x. I will admit that figuring out the units of x is far from easy, so while starting out you might find it easier to assume that the units of x are whatever the denominator of the units of the Average Rate of Change are. So, since in Example 5.2.5 the Average Rate of Change was in wolves per month, then x had to be in months.

Let's wrap up this Section then with a couple more Examples now.

---

[72] This is because of how addition works when you add units.
[73] This is a very tough step to follow. You may wish to read the paragraph after the Example for more on this before finishing the Example.

©Brian Gillispie, 2016.

Example 5.2.6: A lawnmower is able to drive 16 mph when the blades are not cutting grass, and the speed of the lawnmower is reduced by 1.4 mph per inch of grass being cut. How fast can the lawnmower go if the grass is 0.75 inches high?

Solution to 5.2.6: First we need to figure out what is the slope and y intercept for this problem. The slope is the rate of change, which is -1.4 mph per inch of grass[74]. To find the y intercept, notice that we know that the lawnmower goes 16 mph when there is no grass or zero grass, which means 16 is the initial value. Therefore, if we let 16 be the y intercept and -1.4 be the slope, the linear model for this situation is then the following:

$$y = -1.4x + 16$$

Next, use the model to answer the question, which is, how fast can the lawnmower go if the grass is 0.75 inches high? Since y is in the same units as the y intercept, this means y represents the speed of the lawnmower, in mph. Also, since the rate of change was in mph per inch of grass, that means x then is the height of the grass. Therefore, we need to plug in 0.75 for x. Doing that gives us:

$$y = -1.4(0.75) + 16 = 14.95 \text{ mph}$$

Therefore, we would conclude that the lawnmower can go 14.95 mph when the grass is 0.75 inches tall.

Example 5.2.7: Due to heavy rainfall, a river is currently 12 feet high and rising by 0.5 feet per hour. After how many hours will the river be 15 feet high?

Solution to 5.2.7: Like before, we will find the slope and y intercept first. In this problem, we know that the river is changing by 0.5 feet per hour, so our slope is 0.5 feet per hour. Also, we know that the river is 12 feet high, so our y intercept should be 12 then. Plugging both of these into our linear model gives us:

---

[74] The slope is negative because the speed is decreasing as the grass is higher.

©Brian Gillispie, 2016.

$$y = 0.5x + 12$$

Now, we need to use the model to answer the question. This time we wish to know how long until the river is 15 feet high. Since y has the same unit as the y intercept, y is the height of the river. Also, since the rate of change was in the units feet per hour, x has to then be how many hours have passed. Therefore, we need to plug in 15 for y and solve for x. If we do that, we end up with:

$$15 = 0.5x + 12$$

Or, after we solve for y:

$$y = 6$$

Therefore, after 6 hours the river will be 15 feet high.

Before we wrap up this Section, I wish to mention one thing. A common question when working these problems is how to know when to solve for x and when to solve for y? Since most modeling involves quantities and time, and time is always x in these problems[75], there is an easy trick you can use. If the problem asks you how long until something happens, you are solving for x, and that means you plug in your known data into y. However, if the question asks you how much or how many or any other way of asking you the quantity of something, then you are solving for y, and you plug in your given information for x. This will work as long as x represents time. If x does not represent time, then there is no easy way to tell, and instead you'll have to look closer at what units everything is in to tell which one you are now solving for.

---

[75] See Chapter 2.

©Brian Gillispie, 2016.

5.2 Exercises

For Exercises 1 - 6, create a Linear Model given the following information.

1. A deer population has 52 deer when first observed, and is increasing by 3 deer per month.

2. A fruit fly population has 1252 flies when first observed, and is decreasing by 10 flies per month.

3. A car is moving at 15 mph and is accelerating at 2 mph per second.

4. The temperature outside was 52 degrees at 6 pm and is decreasing by 0.5 degrees per hour.

5. There are currently 37500 birds on an island, and the number of birds is increasing by 5000 per year.

6. The cost of gas is $3 per gallon in January 2014, and is increasing by $0.25 per gallon per month.

For Exercises 7 - 17, create a Linear Model, and use your model to answer the question asked.

7. Digital sales of the newest song are initially 4000 sales, and the number of sales is increasing at the rate of 500 per day. How many copies of the song will sell 12 days from now?

8. The cost of milk is currently $2 a gallon, and is increasing by $0.12 per year. How much will milk cost 15 years from now?

9. An island has 100 deer, and the number of deer on the island is increasing by 10 deer a year. How many deer will be on the island 4 years from now?

©Brian Gillispie, 2016.

10. A river has 15000 salmon, and the number of salmon is increasing by 1000 per month. How many months until there are 17000 salmon in the river?

11. A river is 12 feet high, and the height of the river is increasing by 1.2 feet per day. How high will the river be 4 days from now?

12. A river is 12 feet high, and the height of the river is increasing by 1.2 feet per day. After how many days will the river be 16 feet high?

13. Currently it is 56 degrees, and the temperature is dropping by 0.015 degrees per minute. What will the temperature be 72 minutes from now?

14. A snowplow can go 55 mph on a clear road, and for each inch of snow that is on the ground the top speed of the snowplow will drop by 12 mph. How fast can the snowplow go if there is 3.4 inches of snow on the road?

15. A snowplow can go 60 mph on a clear road, and for each inch of snow that is on the ground the top speed of the snowplow will drop by 9.8 mph. Once the snowplow cannot go over 10 mph, it is pulled off the road and a high powered snowplow is sent out instead. How many inches of snow must be on the road at minimum for the high powered snowplow to be sent out?

16. A cafeteria has 2500 pounds of hamburger meat on hand, but is consuming 200 pounds of hamburger meat per day. The next truck delivering meat arrives in 14 days. Does the cafeteria run out of meat? Why or why not?

17. The temperature outside was 25 degrees Celsius at 7 am, and is increasing by 0.9 degrees Celsius per hour. How warm will it be outside at 1 pm?

5.3: Linear Models from Two Data Points

In the last Section, we demonstrated how to create a linear model when one knows the rate of increase or decrease, and one knows the starting value. In this Section, we are going to discuss how to create a linear model with two data points. If we know two data points of the situation we intend to model, we can directly calculate the rate of change and the y intercept ourselves, and then create a linear model from that. However, when creating a linear model this way, you will have to decide on which numerical scale to use, which will affect how your model is used in the end. Let's demonstrate all of this in an Example now.

Example 5.3.1: In the month of May there were 35 owls present, and in the month of June there were 38 owls present. Predict how many owls will be present in August using a linear model.

Solution to 5.3.1: To begin, we need to figure out what x and y represent in our model. Since we have months and owls provided, and time is always x, then x will represent the month. This means that y will then represent the number of owls present in each month.

Next, we need to find our data points. Based on our decisions for what x and y are, our data points are (May, 35) and (June, 38). However, we cannot calculate a slope with those points, so we need to convert May and June to numbers. We will assign to each month the number representing which month in the year it is, which means that x will mean the numerical representation of each month. This means our data points are now $(5, 35)$ and $(6, 38)$

Now, plug our data points into the slope equation. Doing that gives us the following:

$$\frac{38 - 35}{6 - 5} = 3 \text{ owls/month}$$

Next, use the point slope form of the equation of a line to create the equation of a line. Doing that gives us the following equation of a line:

$$y - 35 = 3(x - 5)$$

©Brian Gillispie, 2016.

Or, if we choose to solve for y, we end up with:

$$y = 3x + 20$$

Finally, since x represents the numerical representation of the month, and July is the seventh month, we plug in 7 for x, which gives us:

$$y = 3(7) + 20 = 41$$

Therefore, we conclude that in the month of July there will be 41 owls present.

The reader might notice that you could also calculate the slope, then use the methods of either Section 5.2 or Chapter 3 to solve the problem as well. That works well when the data points are set up in such a way that we can advance forward easily to our desired point in time. However, that is slow for when you need to go many steps into the future, as you probably noticed in Chapter 3 when we wanted to go ten or twelve months forward in time. Also, the methods of the other sections don't work well for problems like this one that we will work next.

Example 5.3.2: During a heavy snowstorm, it is estimated that 1.2 inches of snow had fallen by 1 pm, and 6.0 inches of snow had fallen by 6 pm. How much snow had fallen by 8 pm, using a linear model?

Solution to 5.3.2: First, decide on what the variables represent in the model. Since we have time in the model, x will represent time, in hours after 1 pm. Therefore, this means that y will then represent the amount of snow that has fallen, in inches.

Next, we need to find our data points. Based on our decisions for x and y, our first data point is $(0, 1.2)$, since an x is hours after 1 pm. Our second data point will then be $(5, 6.0)$, as 6 pm is 5 hours after 1 pm[76].

---

[76] If that seems confusing, you technically could make the points $(13, 1.2)$ and $(18, 6.0)$ instead, using military time. In fact, many do just that, as they find the model more straightforward to create and understand this way.

©Brian Gillispie, 2016.

With the points established, we then plug these points into the slope equation to find that our slope is:

$$m = \frac{6.0 - 1.2}{5 - 0} = \frac{4.8}{5} = 0.96$$

With the slope found, plug everything into the point-slope equation for a line. If we do that, we end up with:

$$y - 1.2 = 0.96(x - 0)$$

Which, if we solve for y, we end up with:

$$y = 0.96x + 1.2$$

Finally, we use this model to answer the desired question. We wish to learn how much snow had fallen by 8 pm, and since 8 pm is 7 hours after 1 pm, we plug in $x = 7$ into the equation to end up with:

$$y = 0.96(7) + 1.2 = 7.92$$

Therefore, we conclude that by 8 pm, 7.92 inches of snow had fallen.

Notice that in the last Example, we had to decide on what the numerical scale was for x in our problem. This is normal when using linear models with two points. When that happens, most people choose to let a x value of zero represent the first known data point in the problem. However, by doing this, you will have to always remember what x represents[77]. Let's demonstrate this again on another Example.

---

[77] Recall Section 2.3, when we worked with scaled models. That what we are doing here, working with a scaled model, though this time the scale is one of our choosing.

©Brian Gillispie, 2016.

Example 5.3.3: The high temperature is 89 degrees outside on May 29th, and 90 degrees outside on May 30th. Using a linear model, predict on which day the high temperature will be 93 degrees outside.

Solution to 5.3.3: First, we must decide what x and y represent in this problem. Since we have data on high temperatures and days, and since days is a unit of time and time is always x, we will let x be the date. Therefore, y will be the high temperature for that date[78].

Next, we need to find our data points. Based on our decisions for x and y, our first data point is (May 29th, 89), and our second data point is (May 30th, 90). Since we cannot do math with dates, we need to assign numbers to them. Using our rules from Chapter 2, we will assign May 29th the number 0 and May 30th the day 1, and x now means the number of days after May 29th.

Therefore, our points are now $(0, 89)$ and $(1, 90)$, which, if we plug them into the slope equation we get:

$$m = \frac{90 - 89}{1 - 0} = \frac{1}{1} = 1$$

With the slope found, plug everything into the point-slope equation for a line. If we do that, we end up with:

$$y - 89 = 1(x - 0)$$

Which, if we solve for y, we end up with:

$$y = x + 89$$

---

[78] Note that you can technically reverse this if you wish, and let y be date and x be temperature. It's just not recommended.

©Brian Gillispie, 2016.

Finally, we use this model to answer the desired question. We wish to learn on what day will the high temperature reach 93 degrees? Since y represents the high temperature, that means we need to plug in 93 for y and solve for x. Doing that gives us the following equation:

$$93 = x + 89$$

Which, if we solve for x, we end up with:

$$x = 4$$

Therefore, we conclude that 4 days after May 29th[79], the high temperature will be 93 degrees, using a linear model.

As long as one is careful with setting up the points, the linear model works great for short term predictions. However, because we are using a line when we make a linear model, and lines have as their domain all real numbers, nothing stops someone from using a snow forecast model and ending up with crazy answers like there should be 17,000 inches of snow on the ground by now, or the outside temperature will be 355 degrees Fahrenheit on June 16th. In later Chapters we will discuss how to adjust the model to add a domain restriction to avoid these kinds of outputs from occurring, but until then, use your judgment. If an answer seems unreasonable, chances are then the linear model should not have been used for that situation.

---

[79] The exact date would be June 2nd.

©Brian Gillispie, 2016.

5.3 Exercises

1. The temperature outside was 25 degrees at noon, and 33 degrees at 2 pm. Using a linear model, how warm will it be at 6 pm?

2. A business has sold 4,250 shoes in the month of June, and 4,500 shoes in the month of July. Using a linear model, how many shoes will they sell in September?

3. College tuition at a small school was $6,000 a year in 2006, and it is now $9,000 a year in 2016. Using a linear model, how much will college tuition cost (per year) in 2030?

4. Back in 1990, a popular new video game cost $39.99. In 2015, the popular new video game costs $59.99. Using a linear model, predict how much the popular new game will cost in 2020?

5. Video game prices are often reported rounded to the nearest $5, then a penny is removed from the rounded price, and that is what it is sold for. So for instance, if the video game should cost $48.19 by a model, the price is rounded to $50 (the closest $5 to $48.19), then a penny is removed, and it is sold for $49.99. Using this method, what should the popular new video game cost in 2020, based on your answer in Exercise 4?

6. Using your model in Exercise 4, in which year will the popular new video games cost $79.99?

7. Milk costs are $3 per gallon in March, and $3.50 per gallon in July. Predict how much milk will cost (per gallon) in September.

8. Thanks to hyperinflation, a bottled water that cost $2 on February 3rd now costs $32 on February 6th. Using a linear model, how much will this same bottled water cost on February 14th?

©Brian Gillispie, 2016.

9. The temperature at sea level is currently 36 degrees Celsius, but the temperature one mile above sea level is currently 20 degrees Celsius. Using a linear model, predict what the temperature would be 3 miles above sea level?

10. Using your model in Exercise 9, predict what the temperature would be at 5 miles above sea level, or about the height of Mt. Everest.

11. An island has 50 wolves in the year 2010, and 49 wolves in the year 2011. Use a linear model to predict how many wolves there will be on the island in the year 2015.

12. Use your model in Exercise 12 to predict in what year the wolves will be extinct on the island.

13. You estimate that it takes you 4 hours to pant a 200 square foot wall, and 6 hours to paint a 300 square foot wall. Using a linear model, estimate how long it will take you to paint a 600 square foot wall.

14. If it takes you 2 hours to assemble 15 widgets, and 4 hours to assemble 30 widgets, how long will it take you to assemble 50 widgets, assuming a linear model?

15. The depth of the local lake is 672 feet on June 11th, and 680 feet on June 21st. The lake will flood over its' banks when it hits a depth of 704 feet. On what day will the lake flood, assuming a linear model (and nothing else changing)?

16. A snowstorm has put 1 inch of snow on the ground as of 1 pm, and 7 inches of snow on the ground as of 8 pm. Use a linear model to predict how much snow will be on the ground by 11 pm?

17. The outside temperature is 7 degrees at 1 am and 4 degrees at 2 am. Using a linear model, predict the outside temperature at 6 am.

©Brian Gillispie, 2016.

5.4: Applications of Linear Models

Over the course of the last few Sections, we have discussed how to create linear models from the given information. Now we are going to demonstrate two well-known linear models and show how they are used, the cost/revenue models, and the supply/demand models.

5.4.1: Cost and Revenue Models

When running a business, you have costs that are associated with running the business, as you have to pay the rent, pay all the employees, pay the insurance, pay for the equipment, and so on. Some of these costs are fixed and do not change over the short term, and some of the costs vary based on how much of a product your business made in that time period. As a result, for short-term costs, a business can break up its expenses into what are known as fixed costs and variable costs, which are defined as follows:

Definition 5.4.1: The **fixed cost** of a business are any costs that do not change over the short term, and are therefore not dependent on how much of a product was made

Definition 5.4.2: The **variable cost** of a business is any cost that is dependent on how much of a product was made.

An Example of a fixed cost would be rent on a building. Chances are, your business has already committed to paying that rent, no matter what happens. This means that even if you decide to have your workers take the entire month off, you still owe that rent. Therefore, rent would be considered a fixed cost, at least in the short term.

An example of a variable cost would be paper. If every time your company makes a product, one piece of paper is spent, then the cost of that piece of paper would be a variable cost.

©Brian Gillispie, 2016.

Variable costs are usually the costs of materials involved with making a product, when applicable.

Now that we know how to define fixed and variable costs, let's see how we can put those into an equation. Since the fixed costs never change, the fixed costs will be the same, even if we make zero units of a product. This means then, by the methods of Section 5.2, that the fixed costs will be represented by the y intercept of the model. Also, since variable costs are (usually) the same per item made, at least in the short term, the variable costs can then be represented by the slope of a line. This gives us the following formula for creating a cost equation:

Definition 5.4.3: The **cost equation** for a business for short term operations can be defined as:

$$C = mx + b$$

Where:

$$m = variable\ cost$$
$$b = fixed\ cost$$

Let's now demonstrate how to use this to create a few cost equations:

Example 5.4.1: A business that makes computers has a fixed cost of $1000 per day, and a variable cost of $200 per computer it makes. Create the cost equation for this situation.

Solution to 5.4.1: Since we are using a preexisting equation, we do not need to figure out what x and y mean, as that is given in the equation. Therefore, since our fixed cost is given as $1000, that is the y intercept in the cost equation. Also, since our variable cost is given as $200, that is the slope in the cost equation. Plugging those into our given equation gives us:

$$C = 200x + 1000$$

©Brian Gillispie, 2016.

Example 5.4.2: A bee farm has fixed costs of $50 per day, and variable costs of $1 per jar of honey it makes. Predict how much it will cost the farm to make 25 jars of honey.

Solution to 5.4.2: To begin this problem, we need to create the cost equation. We know that our fixed cost is $50, so that is the y intercept in the cost equation. Also, we know that our variable cost is $1, so that is the slope in the cost equation. Plugging all of this into our given equation gives us:

$$C = x + 50$$

Next, we wish to predict how much it will cost us to make 25 jars. To find this answer, plug 25 into the equation for x to get:

$$C = 25 + 50$$

Which simplifies to give us that it will cost the business $75 to make 25 jars of honey a day.

Oftentimes, the business is interested in the amount it needs to make to break even. In order to do this, we need to know how much revenue the business brings in. Our revenue equation is as follows:

Definition 5.4.4: The **revenue equation** R, when an object can be sold for m dollars per object, is given by:

$$R = mx$$

This equation will work for our purposes, but does have the disadvantage that it makes no allowance for how much sales will drop as prices are raised[80]. For that reason, it is usually

---

[80] Modifying the equation to account for demand would change the equation into a quadratic, which requires future Sections of the book to be able to solve.

©Brian Gillispie, 2016.

assumed that the price the object is sold at is fixed, and was determined in advance to be the optimal price for supply and demand reasons.

With the revenue equation now defined, let's now define what a break-even point is:

> *Definition 5.4.5:* The **break-even** point for a business occurs when revenue equals cost, or:
>
> $$R = C$$

In other words, to find a break-even point, one must set the revenue and cost equations equal to each other. Sometimes, this requires finding the revenue and cost equations first, which we will now demonstrate with an Example.

Example 5.4.3: A company that makes calculators finds that its daily fixed costs are $500, and it costs $15 to make each calculator. Each calculator can be sold for $35 per calculator. How many calculators does the business have to sell to break even?

Solution to 5.4.3: Begin this problem by creating the cost and revenue equations. The fixed cost is $500, so that is the y intercept in the cost equation. The variable cost is $15, so that is the slope in the cost equation. Also, since calculators can be sold for $35 each, that is the slope in the revenue equation. This gives us the following two equations:

$$C = 15x + 500$$
$$R = 35x$$

To find the break-even point, set the revenue and cost equations equal to each other, which results in the following equation:

$$15x + 500 = 35x$$

©Brian Gillispie, 2016.

Solve for x to find that the break-even point is 25. Therefore, if the business sells 25 calculators, they will break even for the day.

The cost and revenue equations can be used to solve for missing fixed and variable costs, or even solve for the price per item sold, as we will demonstrate now in these last two Examples.

Example 5.4.4: A phone company can make phones for a fixed cost of $1,200,000 and a variable cost of $10 per phone. If market research determines that they will sell 75,000 phones, what is the minimum they can charge for the phone and still break even?

Solution to 5.4.4: Begin this problem by creating the cost and revenue equations. The fixed cost is $1,200,000, so that is the y intercept in the cost equation. The variable cost is $10, so that is the slope in the cost equation. This time we don't know how much we can sell each phone for, but we do know we will sell 75,000 phones. Since x stands for the number of items sold, plug in 75,000 for x into both equations. This makes our two equations:

$$C = 10(75000) + 1200000$$
$$R = m(75000)$$

Now, since we want to know the least we can sell those phones for to break even, we need to set the two equations equal to each other. Doing that gives us the following:

$$10(75000) + 1200000 = m(75000)$$

Solve this equation for m to get that the company will break even provided m is 26. Therefore, the company needs to sell the phones for at least $26 to break even.

Example 5.4.5: A company recently reported a loss at their shareholder meeting on their sales of the new tablet. Given that the variable costs for the tablet were $76, and the tablet sold for $199, and they sold 500,000 units, what is the highest the fixed costs were for the company to still have reported a loss?

©Brian Gillispie, 2016.

Solution to 5.4.5: Begin this problem by creating the cost and revenue equations. The fixed cost is unknown, so leave it as b in the equation. The variable cost is $76, so that is the slope in the cost equation. Each tablet sold for $199, so that is the slope in the revenue equation. And finally, since we know they sold 500,000 units, that mean we know x, so plug that in for x. Doing all of that yields the following cost and revenue equations:

$$C = 76(500000) + b$$
$$R = 199(500000)$$

Since we wish to know the highest the fixed cost could be to have a loss reported, that means we first need to know the fixed cost required to break even. Then, any cost above that means a loss will occur, as a higher fixed cost will raise the total cost, resulting in the cost being higher than the revenue. Therefore, to find the fixed cost required to break even, set the two equations equal, which results in the following equation:

$$76(500000) + b = 199(500000)$$

Solve the resulting equation for b to find that b is $61,500,000. Therefore, the fixed costs to make that tablet had to be more than $61.5 million dollars for this loss to have occurred.

©Brian Gillispie, 2016.

5.4.2: Supply and Demand Models

Pretend you own a business, and you wish to sell tomatoes. Based on your past experience, you know that if you sell the tomatoes for $3 a pound, you will sell 2000 pounds of tomatoes, but if you sell the tomatoes for $1 a pound, you will sell 5000 pounds of tomatoes. This time you intend to sell the tomatoes for $2 a pound. How many tomatoes can you sell at that price?

This problem here is a classic supply and demand problem. In this situation, you have an object (tomatoes) that you wish to sell, and you want to know the demand for the object. You know that if you charge too much for the tomatoes, no one will buy them from you, and if you charge too little, everyone will be lining up to buy your tomatoes. Turns out that since we have two data points, we can model this situation by a linear demand equation, which looks like the following:

$$D = mp + b$$

In this equation, p is the price charged for the object for the object, and D is the quantity demanded[81]. Once this demand equation is known, a business can use it to predict demand in new situations, which we will demonstrate now.

Example 5.4.6: A business notices that when they charge $5 for a pound of potatoes, that they sell 1000 pounds in a week, and when they change $3 for a pound of potatoes, they sell 5000 pounds in a week. How many potatoes will be demanded if the business charges $4 for a pound of potatoes?

Solution to 5.4.6: Since the supply and demand equations are used to predict the quantity, given the price, our data points should be of the form (price, quantity). Therefore, that means we have

---

[81] Notice p is where x usually is in the equation, and D is where y usually is in the equation. For that reason, some prefer to still use y = mx+b, with the understanding that x is price, and y is quantity.

©Brian Gillispie, 2016.

the following two data points: $(5, 1000)$ and $(3, 5000)$. With those two points, we can find the slope, which is:

$$m = \frac{5000 - 1000}{3 - 5} = -2000$$

With the slope found, we can create the equation of the line. Using point-slope form, we find that the equation that represents our demand is[82]:

$$D - 1000 = -2000(p - 5)$$

Which, simplifies to give us the following demand equation:

$$D = -2000p + 11000$$

Finally, we wish to use this equation to predict the quantity demanded if the price is $4. Plugging in $4 for p gives us:

$$D = -2000 * 4 + 11000 = 3000$$

Therefore, if the business decides to charge $4 a pound of potatoes, they can expect about 3000 pounds of potatoes to be demanded by customers.

Similar reasoning can be used to create a supply equation for a business as well. If a business has two past data points, then the business can create the following supply equation:

$$S = mp + b$$

Where S is the quantity supplied, and p is the price per item. Let's demonstrate the supply equation now.

---

[82] Remember, p took the place of x and D took the place of y in the equation, so p will also take the place of x and D will also take the place of y in the point-slope equation as well!

©Brian Gillispie, 2016.

Example 5.4.7: The same business also notices that if they pay suppliers $2 a pound, only 5000 pounds of potatoes arrive, and if they pay suppliers $3 a pound, only 7000 pounds of potatoes arrive. How many pounds of potatoes can they expect to be supplied if they offer to pay suppliers $3.50 a pound?

Solution to 5.4.7: Since the supply equation is used to predict the quantity, given the price, our points should be of the form (price, quantity). Therefore, our data points should be $(2, 5000)$ and $(4, 7000)$. Using those two points, we calculate our slope, which is:

$$m = \frac{7000 - 5000}{4 - 2} = 1000$$

With the slope found, use point slope form to create the equation of the line, which gives us:

$$S - 5000 = 1000(p - 2)$$

Which, simplifies to the following equation:

$$S = 1000p + 3000$$

Finally, plug in 3.5 for p to answer the question, and we end up with:

$$S = 1000 * 3.5 + 3000 = 6500$$

Therefore, if we offer to pay businesses $3.50 per pound of potatoes, we should expect about 6500 pounds of potatoes to be supplied to us.

In the previous Examples, notice that the supply curve has a positive slope, and the demand curve has a negative slope. This makes logical sense, because a positive slope on the supply curve means that if you pay people more to provide you with a product, then more of that

©Brian Gillispie, 2016.

product will be provided. If someone decided to post an ad saying they would pay $100 per tomato, the line of people wishing to sell them tomatoes would probably be miles long! However, if you posted an ad saying you would pay someone $0.01 per tomato, you would, most likely, have no one take you up on your offer. The positive slope on the supply curve exists because as we raise the price we pay people, more people will want to supply us with the object that we wish to buy.

The demand curve having a negative slope also makes practical sense. As prices rise, less people decide to buy the object, either due to not being able to afford it, or due to finding other items to buy in its place (economics call this finding a substitute). If someone decides to charge $200,000 for a ticket to a popular football game, chances are good that no one will buy it. However, if someone charges $2 for a ticket to a popular football game, then the line of people wanting to buy those tickets would be really long. However, there is a finite limit to how many people will still buy the object, even if you give it away. This is represented by the y-intercept of the demand curve.

To see how the market works with these supply and demand curves, we are going to pretend that there is no markup between the supplier and the resold price. In other words, businesses will sell all items at the same price they paid their suppliers. This simplifying assumption will allow us to still gleam some insights into how the market works, as we will show now.

Example 5.4.8: Using the following supply and demand equations for pounds of potatoes:

$$D = -2000p + 11000$$
$$S = 1000p + 3000$$

Predict how many pounds of potatoes will be supplied and demanded when the price is $2 per pound.

Solution to 5.4.8: To find the answer to this, we need to plug in 2 for p into both the supply and demand equation. Doing that gives us:

$$D = -2000(2) + 11000 = 7000$$
$$S = 1000(2) + 3000 = 5000$$

This means that at $2 a pound, the demand is for 7000 pounds of potatoes, but the supply will only be 5000 pounds of potatoes.

In the scenario in Example 5.4.8, the company is being supplied less potatoes than are being demanded by customers. When this happens, a shortage occurs. Usually when a shortage occurs in the market, the company will raise the price they are selling the object for, in order to take advantage of the shortage. So, if the situation presented in Example 5.4.8 occurred for a real business, they would most likely raise the price on the potatoes to $3 a pound, in order to make more money, and in order to slow down the demand so they only sell what they have available.

Example 5.4.9: Using the same supply and demand equations from Example 5.4.8, find the pounds of potatoes supplied and demanded if the price is $5.

Solution to 5.4.9: In this case, we need to plug in 5 for the price. Doing this gives us:

$$D = -2000(5) + 11000 = 1000$$
$$S = 1000(5) + 3000 = 8000$$

Which means that if the business charges $5 a pound, they will be supplied 8000 pounds of potatoes, but the demand is for 1000 pounds of potatoes.

In the situation in Example 5.4.9, the business this time is being supplied more potatoes than they have demand for. When this happens, they have a surplus. Now, the business still needs to get rid of the extra potatoes, as it does them no good for the potatoes to go bad. Therefore, what usually happens in this situation is the business will lower the price they are

©Brian Gillispie, 2016.

charging on the potatoes, in order to sell off all the potatoes they were supplied. This is why you often see bakeries holding sales on day old bread. They amount they were supplied by the chefs exceeded demand, and the sale is an attempt to still make something on the product instead of throwing it out.

The supply and demand models, as stated, do teach us something about how the market works. For one, if a business is supplied with more product than the demand is, they will be forced to do something with this extra inventory, and this something is usually a discounted sale. Also, if a business is supplied with less of a product than the demand is, they can (and often do) raise prices so as to try to only sell the amount they were supplied. This is why you will sometimes see high housing prices in a college town. The amount of housing available is less than the amount of housing demanded, so the landlords raise their prices until the amount supplied equals the amount demanded. To see this in action in the real world, look up the housing prices in Watford City, North Dakota [9] over the last few years[83]. Watford City has been the location of oil projects over the last few years, and as a result, many people have been moving to the town to work. However, the influx of people has overloaded the available housing, and as a result, rent in that town has jumped from $500 a month to $1500 a month in most areas over the last few years! This jump is due to demand outpacing supply, and will only drop when more housing is built and available in the area (which is starting to happen now, see [10]).

---

[83] As of the time this paragraph was written that is, so the housing prices are from early 2014.

©Brian Gillispie, 2016.

5.4 Exercises

Exercises that begin with F require the material in Section 5.4.1, and Exercises that begin with S require the material in Section 5.4.2.

F1. A business that makes computers has a fixed cost of 100,000 per month, and a variable cost of 250 per computer. If each computer sells for 750 per computer, how many do they need to sell to break even?

F2: A farmer decides to grow corn for the year. The yearly fixed costs for the field is $1000, and each pound of corn costs the farmer $1 to grow. If the farmer can sell the corn for $2 per pound, how many pounds does he need to sell to break even?

F3: A book writer finds that in the course of writing one book he has a fixed cost of $50000, and each copy of the book printed costs him $4 per book. If each copy of the book is sold for $8, how many copies does he need to sell to break even?

F4: A stadium has a fixed cost of $250,000 for the football season, and each ticket sold incurs a variable cost of $25 per person to the stadium. If each ticket is sold at $75 per ticket, how many tickets does the stadium need to sell for the season to break even?

F5: A company releases a new video game. The fixed costs to make the game were $100,000, but the game has no variable costs associated with it. If each copy of the game sells for $50, how many copies does the company need to sell to break even?

F6: A company finds that it has a variable cost of $5 per item made, and each item sells for $9 per item. The company also finds that they usually sell 50 of these items a day. What is the highest the fixed cost can be for the company to still make a profit?

©Brian Gillispie, 2016.

F7: A company that makes computers has a fixed cost of $1000 per day, and each computer sells for $500 each. If the company sells on average 5 computers a day, what is the highest the variable cost can be, per computer, before the company starts losing money per day?

F8: You decide to make a hit new game and sell it on the internet! Given that you plan to sell your game for $4.99 per copy, and the variable costs for selling your game is $2.49 per copy sold, if market research states that you can expect to sell 5,600 copies[84] total, what is the highest your fixed costs can be for you to still make a profit?

F9: You wish to make a fancy phone that will dominate in the sales market! Given that the fixed costs to make such a phone are $50 million, each phone is going to sell for $399, and market research says you can expect to sell 1.5 million copies, what is the highest your variable costs can be and you still make a profit on this phone?

F10: You have designed the newest hit video game console! If the console is going to sell for $499 each, market research says you can expect to sell 6 million copies, and your variable cost for each console sold is $399, what is the highest the fixed costs can be for you to still make a profit?

F11: At a shareholder meeting, someone let slip that they needed to sell 13.5 million copies of their newest hit video game or they will take a loss on the making of it. If each copy of the game sells for $70, and the variable costs for each copy sold is $49, what were the fixed costs for making this game?

F12: You plan to release a new TV model into the market. If your variable costs to make the TV are $125 per TV, and the fixed costs are 5.5 million, what is the lowest you can sell the TV for and make a profit, if market research says you can expect to sell 575,000 TV's?

---

[84] As of the writing of this problem (2016) that is actually above the average number of copies the average game sells.

©Brian Gillispie, 2016.

S1: If a business charges $10 per CD, they find they sell 10,000 of them, and if they charge $12 per CD, they find they sell none of them. Create the demand equation for this situation.

S2: If a business charges $10 per baseball game, they find they sell 12,500 tickets, and if they charge $20 per baseball game, they find they sell 2,500 tickets. Create the demand equation for this situation.

S3: The stadium in question in Exercise S2 sells out if they ever sell 40,000 tickets. Based on this model, will the stadium ever sell out a baseball game?

S4: What are some of the factors the models in Exercise S2 and S3 are not accounting for? Based on these missing factors, do you think the stadium would ever sell out a game now?

S5: A business finds that if they pay suppliers $2 per pound of onions, they receive 10,000 pounds of onions, and if they pay suppliers $1.75 per pound of onions, they receive 8,000 pounds of onions. Create the supply equation for this situation.

S6: Use your supply equation in Exercise S5 to predict how many pounds they will be supplied if they pay suppliers $1.50 a pound.

S7: A grocery store finds that if they charge $5 per pound of hamburger, then they sell 100 pounds a day, and if they charge $4 per pound of hamburger, they sell 200 pounds a day. Also, if they pay suppliers $5 per pound, they receive 200 pounds of hamburger, and if they pay suppliers $4 per pound, they receive 125 pounds of hamburger. Create the supply and demand equations for this situation.

S8: Using the equations in Exercise S7, how many pounds of hamburger will be supplied and demanded if the price is $4.50 a pound? Is this a shortage, or a surplus?

S9: Using the equations in Exercise S7, how many pounds of hamburger will be supplied and demanded if the price is $3 per pound? Is this a shortage, or a surplus?

©Brian Gillispie, 2016.

S10: For the equations in Exercise S7, at what price will the store have no pounds of hamburger supplied?

S11: For the equations in Exercise S7, at what price will the store sell no pounds of hamburger?

S12: A grocery store finds that if they charge $2 for a box of doughnuts, they sell 11,000 boxes a day, and if they charge $10 for a box of doughnuts, they sell 3,000 boxes a day. Similarly, if they pay suppliers $2 to make a box of doughnuts, only 1000 boxes are supplied, and if they pay suppliers $10 to make a box of doughnuts, 8000 boxes are supplied. Create the supply and demand equations for this situation.

S13. For the equations in Exercise S12, how many boxes will be supplied and demanded if the price is $8 a box. Is this a shortage or a surplus?

S14: For the equations in Exercise S12, how many boxes will be supplied and demanded if the price is $5 a box. Is this a shortage or a surplus?

S15*: Usually the price for the demand equation and the price for the supply equation are not the same, as the store adds a mark-up to the price. For the equations in Exercise S12, how many boxes are supplied and demanded if the price for suppliers is $5, and the price the boxes are sold at is $9.50? Do you have a shortage or a surplus?

©Brian Gillispie, 2016.

Chapter 5 Summary

5.1: In this Section, we covered how to find the equation of a line, given two points. If we have two points (x_1, y_1) and (x_2, y_2) then we can find the slope by computing:

$$m = \frac{y_2 - y_1}{x_2 - x_1}$$

Once we have the slope, we can create the equation of the line by using either point-slope form or slope-intercept form. Point-slope form is:

$$y - y_1 = m(x - x_1)$$

And, slope-intercept form is, with b meaning the y-intercept:

$$y = mx + b$$

5.2: In this Section, we discussed how to interpret the slope and the y intercept. The slope stands for the rate of increase or decrease of the item being modeled, and the y intercept stands for the starting amount or initial amount of the object being modeled.

5.3: In this Section, we discussed how to create linear models when two points are given. If we know the model needs to contain the two points (x_1, y_1) and (x_2, y_2), then a linear model can be used to model this situation.

©Brian Gillispie, 2016.

5.4: In this Section, we discussed two applications of linear models. In the first application, we saw that a short-term cost equation can be given by the following, with m representing the variable cost, and b the fixed cost:

$$C = mx + b$$

Also, if we know how much we make per item sold, then a revenue equation can be found by the following, with m being the cost each item is sold at:

$$R = mx$$

In our second application, we looked at how we can use a linear equation to represent supply and demand equations. The linear supply and demand equations are created from two data points each, and can be written as:

$$D = m_1 * p + b_1$$
$$S = m_2 * p + b_2$$

Where m_1 and m_2 are the respective slopes of each line, and b_1 and b_2 are the respective y intercepts of each line.

©Brian Gillispie, 2016.

Chapter 6: Exponential Models

In the last Chapter, we explored and discussed how to use our first continuous model to make predictions about the future. In this Chapter, we will explore and discuss another continuous model, the exponential model.

First, we will begin our discussion by discussing common terms used in exponential models, as well as review how to solve exponential equations. From there, we will spend a few Sections working with our exponential model as we tweak and refine it to fit various situations. Then, we will conclude by looking at a few various well-known exponential models and how they are used.

©Brian Gillispie, 2016.

6.1: Review of Exponents and Logarithms

Before we can begin working with exponential models, we need to review some of the common terms involved when working with exponents, as well as how to solve exponential equations. The reader who is familiar with these topics can safely skip ahead to Section 6.2.

6.1.1: Exponents

Exponents often appear in mathematical equations in the form x^y, which is read x to the power y. In these problems, x is referred to the base and y is referred to as the exponent. So for example, if you have 2^4, then 2 is the base and 4 is the exponent in that statement. To calculate an exponent, remember that the exponent tells you how many times the base appears in the expression, all being multiplied. So for instance, if you have 2^4, this means that to compute it we need to calculate $2 * 2 * 2 * 2$, as the exponent tells us there are 4 2's in the problem, all being multiplied to each other. Let's demonstrate this now in an Example or two before proceeding:

Example 6.1.1: Evaluate 3^5.

Solution to 6.1.1: 3^5 reads three to the fifth power and means $3 * 3 * 3 * 3 * 3$, which calculates to 243. Therefore, $3^5 = 243$

Example 6.1.2: Evaluate 10^4.

Solution to 6.1.2: 10^4 reads ten to the fourth power, and means $10 * 10 * 10 * 10$, which calculates to 10,000. Therefore, $10^4 = 10,000$

©Brian Gillispie, 2016.

Usually we will not compute out exponents by hand but will instead rely on calculators. We will assume that the reader has a calculator that can calculate exponents going forward in all our future calculations[85].

Next, we will review the common rules for exponents, which are as follows:

Rules of Exponents

1. $a^0 = 1$, provided $a \neq 0$
2. $1^x = 1$, for any x
3. $a^m * a^n = a^{m+n}$
4. $\frac{a^m}{a^n} = a^{m-n}$
5. $(a^m)^n = a^{m*n}$
6. $a^{-n} = \frac{1}{a^n}$

We will not directly use all of the rules in this list, but it is still good to be aware of them for the times when they are needed. Let's demonstrate a couple of these rules now before proceeding.

Example 6.1.3: Evaluate 1.05^0

Solution to 6.1.3: By the first rule of exponents, anything to the zero power is 1, so therefore $1.05^0 = 1$

Example 6.1.4: Simplify $(x^2)^5$

Solution to 6.1.4: By rule five of the rules of exponents, we can simplify this by multiplying the exponents. Therefore:

$$(x^2)^5 = x^{2*5} = x^{10}$$

---

[85] Usually one computes an exponent using the carrot key or the x^y button. For more details, please check your calculator owner's manual, as the directions vary by calculator and model.

©Brian Gillispie, 2016.

Next, we wish to define the number e, which is a number that appears in various exponential models. The number e is defined as follows:

> *Definition 6.1.1:* The number e is defined as follows:
>
> $$e \approx 2.71828\ldots$$

While there is a more mathematically precise definition available, that definition will not help us at all when we need to calculate e^3, so the definition given above is the one that we will use. In practice, most people use the e button on their calculator as needed for computations. However, do remember that e is also approximately 2.71828, in case you don't have a calculator that calculates e on hand.

6.1.2: Logarithms and Solving Exponential Equations

In modeling, it is common to need to be able to solve equations of the form $2^x = 10$, where the variable we are solving for is an exponent. In order to solve equations of this time, we will be required to use logarithms. The definition of a logarithm is as follows:

> *Definition 6.1.2:* The **natural logarithm function**, or $ln(x)$ for short, is defined as follows:
>
> $$\ln(a) = b \text{ if and only if } e^b = a$$

For this book, we will not use this definition of a logarithm, as we will be using a calculator to compute all such logarithms when they arise. Instead, we will be using two rules of logarithms in order to solve our equations. The first rule is as follows:

©Brian Gillispie, 2016.

> *Definition 6.1.3:* Natural log and *e* share the following relationship:
>
> $$\ln(e^x) = x \text{ and } e^{\ln(x)} = x$$

Let's demonstrate using this first rule in two Examples before we continue:

Example 6.1.5: Compute $\ln(e^{10})$

Solution to 6.1.5: By Definition 6.1.3, $\ln(e^x) = x$, so then $\ln(e^{10}) = 10$

Example 6.1.6: Compute $e^{\ln(5)}$

Solution to 6.1.6: By Definition 6.1.3, $e^{\ln(x)} = x$, so then $e^{\ln(5)} = 5$

As you can see, this rule allows us to significantly simplify exponential equations. However, the previous rule is dependent on the exponential being *e*. It turns out there is an equivalent version that we can use for exponentials that are not *e*, which we shall define now.

> *Definition 6.1.4:* The natural logarithm function and exponentials are related in the following way:
>
> $$\ln(a^x) = x * \ln(a)$$

Definition 6.1.4 is what we will be using to solve exponential equations when they arise in modeling, as we can legally take the natural log of both sides of the equation, which will allow us to then use this rule. From there, we can use our rules of algebra to finish solving the problem. Let's demonstrate this with an Example now.

©Brian Gillispie, 2016.

Example 6.1.7: Solve for x: $5^x = 75$

Solution to 6.1.7: Since the variable we wish to solve for is an exponent, let's begin by taking the natural log of both sides. Once we do that, we end up with.

$$\ln(5^x) = \ln(75)$$

Now, by Definition 6.1.4, we can re-write $\ln(5^x)$ as $x * \ln(5)$. Substituting $x * \ln(5)$ in for $\ln(5^x)$ gives us:

$$x * \ln(5) = \ln(75)$$

Finish by dividing both sides by $\ln(5)$. If we do that, we end up with:

$$x = \frac{\ln(75)}{\ln(5)}$$

Which, is the final answer[86] to this problem, unless we use a calculator to simplify. Using a calculator gives us $x \approx 2.6826$

Do be very careful when using Definition 6.1.4, as the Definition only applies if the *entire* term is raised to the same power. This means that you cannot use Definition 6.1.4 to simplify $\ln(100 * 1.05^x)$, as the exponent of x is not on both the 1.05 and the 100 in that equation. Only if the same exponent is on both of the numbers can you use this rule. For this reason, it is important to simplify the exponential equations so that the number that is raised to a power is by itself before taking natural log of both sides, else you may find yourself unable to proceed any further.

Let's now demonstrate this on an Example where some simplification is needed before we can take the natural log of both sides.

---

[86] Be careful here! Many are tempted to cancel the ln's, but since ln is a function you cannot cancel those! All you can do to simplify it from here is compute $\ln(75)$ and $\ln(5)$ on a calculator, then divide those two terms.

©Brian Gillispie, 2016.

Example 6.1.8: Solve for x: $100 * (1.05)^x = 250$

Solution to 6.1.8: Be careful here. Many are tempted to take the natural log of both sides as their first step. However, Definition 6.1.4 only applies if the **entire** side of the equation is raised to a power, as we noted above. Instead, we have to simplify the equation by dividing both sides by 100, which yields:

$$(1.05)^x = 2.5$$

Now that the exponent term is by itself on one side of the equation, we can take the natural log of both sides. Doing this gives us:

$$\ln(1.05^x) = \ln(2.5)$$

Now, by Definition 6.1.4, $\ln(1.05^x)$ is the same as $x * \ln(1.05)$. Making that substitution into the equation yields:

$$x * \ln(1.05) = \ln(2.5)$$

Finish solving for x by dividing both sides by $\ln(1.05)$, which results in:

$$x = \frac{\ln(2.5)}{\ln(1.05)} \approx 18.78$$

Now that we have reviewed what exponents are and how to solve exponential equations, we will turn our attention to how to create exponential models in the rest of this Chapter. Please refer back to this Section as needed, as future Sections will assume you know the rules of exponents and logarithms as presented here.

©Brian Gillispie, 2016.

6.1 Exercises

For Exercise 1 - 6, use a calculator to compute the following:

1. 5^7

2. e^3

3. $7^{1.5}$

4. $\ln(1.05)$

5. $\ln(89720)$

6. $\dfrac{\ln(980)}{\ln(2)}$

Solve the following equations for x:

7. $2^x = 8$

8. $4^x = 1024$

9. $6^x = 1000$

10. $2^x = 1{,}000{,}000$

11. $35 * (5)^x = 100$

12. $50000 * (0.95)^x = 25000$

©Brian Gillispie, 2016.

13. $10000 * (1.07)^x = 1{,}000{,}000$

14. $200 * (0.5)^x = 50$

15. $1000 * (1.13)^x = 5000$

16. $30 + 20 * (0.8)^x = 40$

17. $100 + (1.1)^x = 500$

18. $100 + 500 * (0.9)^x = 250$

19. $50 * (0.5)^{(x/10)} = 25$

20. $100 * (2)^{(x/12)} = 350$

21. $65 * (0.75)^{(x/5)} = 40$

22. $325 * (1.1)^{(x/30)} = 700$

23. $125 + 100 * (1.25)^{(x/12)} = 500$

©Brian Gillispie, 2016.

6.2: Creating an Exponential Model

Let's start our discussion of how to create an exponential model by looking at a known exponential model and see if we can figure out how such a model was constructed. In 1859, Thomas Austin released 24 rabbits into the wild of Australia [11] [12]. It is widely believed that we can model the number of rabbits in Australia in the following years by the equation:

$$y = 24 * (6.25)^x$$

What we want to do is figure out a way from this that we can devise an exponential model from this model. Assume that we wish to create an exponential model of the form:

$$y = C * a^x$$

And have it match our known model. How can we do this?

Looking at the two models, we can see that a quick way to create an exponential model which matches our given model is to let C be 24 and a be 6.25. However, in order to create a general exponential model, we need to understand what those numbers mean in our model. Let's start by seeing what happens if we let x be 0 in our model. If we do that, we end up with:

$$y = 24 * 6.25^0$$

Which, simplifies down to 24. Therefore, for our model, since C was 24, then C has to stand for the value of y when x is zero. This is often called the starting value or the initial value, but sometimes will only stand for what y is when x is zero.

To see what a has to represent in our model, now calculate what the model returns when x is 1. If we plug 1 into our rabbit model, we end up with:

$$y = 24 * 6.25^1$$

©Brian Gillispie, 2016.

This simplifies to 150. Now, use the fact that when x was 0, y was 24 and when x was 1, y was 150, to calculate the percent 150 is of 24. If we do that, we find that the percent 150 is of 24 is:

$$\frac{150}{24} * 100 = 625\%$$

That number happens to be 6.25 in percent form. This means that for our exponential model, a stands for the percent the new value is of the old, in decimal form. Let's put all of that together now into our model, which is stated as follows:

Definition 6.2.1: An exponential model is of the form:

$$y = C * a^x$$

Where:

$$C = Starting\ Value\ of\ the\ Model\ (C \neq 0)$$
$$a = The\ percent\ the\ new\ value\ is\ of\ the\ old, in\ decimal\ form$$
$$x = number\ of\ time\ intervals\ that\ have\ passed$$

Let's demonstrate this model with a few Examples now.

Example 6.2.1: An observer of migrating birds notices there are 50 birds, and each week the number of birds is 110% of the number of birds present the previous week. Create an exponential model for this situation.

Solution to 6.2.1: In this Example, 50 is our starting amount, so we will plug in 50 for C. Also, each week the number of birds is 110% the number of birds present the previous week, which means that 1.1 is a in our model. Plugging all of that in gives us the following model.

©Brian Gillispie, 2016.

$$y = 50 * 1.1^x$$

Example 6.2.2: A culture of bacteria starts with 100 cells, and each hour the number of bacteria present is 150% of the amount present the previous hour. Create an exponential model for this situation.

Solution to 6.2.2: In this Example, 100 is our starting amount, so we will plug in 100 for C. Also, each hour there are 150% of the number present the previous hour, so 1.5 is a in the model. Plugging all of that into the model gives us:

$$y = 100 * 1.5^x$$

We can also create exponential models if we are given percent increases or percent decreases instead. To do so, first convert the percent increase or decrease to the appropriate percent of, then proceed from there. Let's demonstrate.

Example 6.2.3: A radioactive substance has 100 mg present, and the number of mg is decreasing by 3% per minute. Create an exponential model for this situation.

Solution to 6.2.3: In this Example, 100 is our starting amount, so we will plug in 100 for C into the model. Also, we know that each minute the number of cells is decreasing by 3% per minute, which, if we convert that to the appropriate percent of, translates to each minute the number of cells is 97% of the previous number of cells. This means that 0.97 is a in the model then. Plugging all of this into the model gives us:

$$y = 100 * 0.97^x$$

Since usually exponential models are used when we know the percent increase or percent decrease, most of our models stated going forward will be in that format. Please make sure you are comfortable converting to a percent of[87] if you are given a percent increase or decrease because going forward we will not be showing the conversion!

Once we have created our model, we can use them to make predictions about the future, by plugging in the appropriate value for x and doing the calculations.

Example 6.2.4: Use the model in Example 6.2.1 to predict how many birds there will be in 10 weeks.

Solution to 6.2.4: In Example 6.2.1, we created the model $y = 50 * 1.1^x$, and our data said that each week there are 110% of the previous week's birds. Therefore, x is in weeks, so we plug in 10 for x, which gives us:

$$y = 50 * 1.1^{10} = 129.69$$

Or, rounding to the nearest bird, 130 birds

Example 6.2.5: Use the model in Example 6.2.1 to predict how long before there is one million birds.

Solution to 6.2.5: To answer this question we need to solve the following exponential equation:

$$50 * 1.1^x = 1,000,000$$

Divide both sides by 50, which give us:

$$1.1^x = 20,000$$

Now, use the rule of logarithms covered in Section 6.1. Take natural logarithm of both sides as follows:

---

[87] If not, this is covered back in Chapter 3.

©Brian Gillispie, 2016.

$$\ln(1.1^x) = \ln(20{,}000)$$

Using the rule of logarithms covered in Section 6.1, re-write the left side of the equals sign as follows:

$$x * \ln(1.1) = \ln(20{,}000)$$

And finally, divide both sides by $\ln(1.1)$ to finish solving for x, giving us the final answer of:

$$x = \frac{\ln(20{,}000)}{\ln(1.1)}$$

Using a calculator to compute the natural logarithms gives us that $x \approx 103.9079$. So if the trend continues, at this rate there will be one million birds in about 104 weeks, rounded to the nearest week.

Be careful when working with this exponential model, as the units of time that x is in is the same units of time the percent is in, whether it is a percent of, percent increase, or percent decrease. This means that if the percent is how much the new value is of the old after 10 years pass, x is now in 10 year increments! Let's demonstrate this with an Example now.

Example 6.2.6: A town is having a population boom! Initially were 1,276 people, and the number of people in the town is increasing by 150% every 10 years. Predict how many people will be in the town 40 years later, using an exponential model.

Solution to 6.2.6: Start off by creating the model. Since we initially have 1276 people, that will be C on our model.. Also, we know that the number of people is increasing by 150% every 10 years, which means that every 10 years, the number of people present is 250% of the previous number of people. This means that a will be 2.5 in our model, but it also means x will mean the

©Brian Gillispie, 2016.

number of 10 year intervals that have passed since our starting time! Plugging all of that in gives us the following model:

$$y = 1276 * 2.5^x$$

Now, to use this to predict the number present 40 years later, we have to be careful, as x is in 10 year intervals. This means that we need to plug into x the number of 10 year intervals that have passed, which means we need to plug in 4 for x, as 40 is 4 10 year intervals. Doing that gives us:

$$y = 1276 * 2.5^4 = 49{,}843.75$$

Or 49,844 if we round to the nearest whole person.

So far every Example has used the word initial or starting value, but sometimes data is presented in another way. When this happens, you have to scale the data in such a way that the time associated with your starting value is represented as an input of 0 in your model. For instance, if you collect data on ducks on a pond in May, you then have to treat an input of 0 into the model as May. This may affect all other values that you plug in as well. Let's demonstrate.

Example 6.2.7: A researcher notices that the population of ducks is decreasing for some reason. In the month of February, there are 120 ducks, and the number of ducks present is decreasing by 40% every month. Predict how many ducks will be present in April, using an exponential model.

Solution to 6.2.7: This time we are given a starting value, but worded differently. Because the only value we know is the number of ducks present in February, that value has to be our starting value, or C. In addition, we know that each month the number of ducks is decreasing by 40%, which means that every month, the number of ducks present is 60% of the previous number of ducks. This means a is 0.6 in our model then. Plugging all of that in gives us the following model:

$$y = 120 * 0.6^x$$

To find out the number of ducks present in April, we need to plug in the appropriate number into our model. While x is in months due to the fact that our percent is in months, we can't plug 4 into the model to get our answer. See, if we plug 4 into the model, the model will return the number of ducks present 4 months after February, because we started the model on the month of February. Therefore, we have to figure out how many months after February April is, and plug that number into the model model. Turns out that April is 2 months after the month of February, which means to answer our desired question we need to plug 2 into the model. Doing that gives us:

$$y = 120 * 0.6^2 = 43.2$$

Therefore, in the month of April there will be 43 ducks, rounding to the nearest duck.

6.2 Exercises

For Exercises 1 - 8, create the exponential model.

1. A cup of coffee is 165 degrees, and the temperature the next minute is 98% the temperature of the previous minute.

2. A culture of bacteria has 250 cells, and the number of cells present the next hour is 200% the number of cells currently present.

3. Ten rabbits escape a ship and end up on a deserted island, and the number of rabbits present the next year is 500% the number of rabbits in the previous year.

4. A dose of medicine has 325 mg of a drug in it, and each hour the percent of the drug present is 75% the amount of the drug previously present.

5. An isolated town has a population boom! In the year 2010 there are 54 people, and the number of people present each year is increasing by 10%.

6. A tree planted in the backyard is 10 feet tall in the year 2019, and the height of the tree is increasing by 25% each year.

7. A pond has 1400 catfish, and the number of catfish is increasing by 6% per month.

8. Due to an accident, 100 grams of Cessium-137 has leaked into the surrounding area, and the amount of Cessium-137 in the area is decreasing by 2.27% per year.

©Brian Gillispie, 2016.

For the rest of the Exercises, create an exponential model, then use it to answer the question asked.

9. A cup of coffee is 165 degrees, and the temperature the next minute is 98% the temperature of the previous minute. How hot is the coffee after 10 minutes pass?

10. Using the model you created in Exercise 9, calculate how long must pass before the coffee is 100 degrees.

11. A drug administers a dose of 500 mg when taken, and the amount of drug still present after an hour is 86% the amount of the drug previously present. How much of the drug will be present after 4 hours?

12. A restaurant has sales of $3,000 on Friday night, and each week the Friday night sales are 104% of the sales of the previous week. What will sales be in 12 weeks?

13. Using the model you created in Exercise 12, calculate how many weeks must pass before Friday night sales are $7,000 or more?

14. A tree is currently 10 feet tall, and every month the height of the tree is increasing by 2%. How tall will the tree be after 1 year passes?

15. A doctor needs to give a patient an antibiotic that has a dose of 1000 mg. After an hour passes, the amount of antibiotic present in the patient has decreased by 6%. The doctor cannot give another dose to the patient safely until the amount of antibiotic present in the patient is 400 mg or less. How much time must pass until the amount of antibiotic present in the patient is 400 mg or less?

16. A drug administers a dose of 650 mg when taken, and the amount of the drug present is decreasing by 44% per 4 hours. How much of the drug will be present after 24 hours pass?

©Brian Gillispie, 2016.

17. A culture of bacteria has 300 cells, and every 6 hours, the number of cells is increasing by 100%. How many cells will be present after 18 hours pass?

18. A pond has a frog explosion! If the pond starts with 58 frogs, and the number of frogs is increasing by 124% every 2 months, how long until there are 280 frogs present?

19. There are currently 26 ducks on a pond, and the number of ducks is increasing by 15% per month. How long until there are 60 ducks?

20. There are currently 150 bacteria cells in a dish, and the number of cells present is increasing by 7% per minute. How many cells are present after 30 minutes?

21. In the year 2010 there are 2500 people in a town, and the number of people is increasing by 1% per year. How many people are present in the year 2016?

22. On October 3rd there are 15,000 birds present at a wildlife refuge, but due to winter migration, the number of birds is decreasing by 4% per day. How many birds are present on October 10th?

23. A pond has 450 catfish in the month of May, and the number of catfish present is increasing by 33% every 2 months. How many catfish will be present in the month of August?

24. Using your model from Exercise 23, predict on which month the pond will have 700 catfish.

6.3: Shifting the Time Interval

The exponential model, in its current form, works well for modeling situations where we know the percent the new value is of the old value. However, the current form of the exponential model also requires that the time interval used for x be the same as the time interval for the percent. This is not a problem if your percent is based on consecutive time intervals, but what if it is based on time intervals that are 2 or 3 units apart? As you saw in the last Section, this can get confusing quick, as it is easy to forget that your units for x are now in 2 day intervals or 6 hour intervals or so on. For this reason, we are interested in finding out if there is a better way to handle this situation. Or, in other words, can we modify the model so as to avoid the problem of having to remember that your inputs are in 2 day intervals or 6 hour intervals or so on. It turns out the answer is yes, which we will demonstrate with an Example now.

Suppose you know there are 150 frogs on a pond, and you also know that the number of frogs on the pond three months from the last measurement is 140% of the number of frogs previously present. By the methods of Section 6.2, we would use the following to model this situation:

$$y = 150 * 1.4^x$$

However, when we wish to use this model, we have to remember that x is in 3 month intervals. So if we wanted to know how many frogs were present in 6 months, we would need to plug in 2 for x instead of 6, as 6 months is 2 3 month intervals.

Now, let's look at how to fix the model so that we could plug 6 in for 6 months instead of 2. To see how to fix the model, we have to look at how we compute that 6 is 2 3 month intervals. When we computed how many 3 month intervals passed in 6 months, we computed:

$$\frac{6}{3}$$

Or, we divided 6, the number of months that have passed, by 3, the length of our intervals. This gave us that 2 3 month intervals had passed, and we plugged 2 into the model.

©Brian Gillispie, 2016.

Could we have built this calculation into the model instead? It turns out the answer is yes. If we had, instead, used the following for our model:

$$y = 150 * 1.4^{x/3}$$

Then we could have plugged in 6 for x now, and got the correct answer still. By adjusting the model in this way, we will no longer have to remember if x is in 3 month intervals, as the model will take care of that. In general, if we know that after d time has passed the value is now $a\%$ of the previous value, we have the following model, given as Definition 6.3.1

Definition 6.3.1: If C is the starting value, and after d time units have passed, there is $a\%$ of the previous value present, then the exponential model for the situation is as follows, with a in decimal form:

$$y = C * a^{(x/d)}$$

Remember that if you have a percent increase or decrease instead to convert it to a percent of first. Let's demonstrate this new model in a few Examples now.

Example 6.3.1: It's a bird invasion! For some reason, a lot of eagles have started arriving in town. In January there are 10 eagles, and every two months there are 150% as many eagles as there were before. Create an exponential model to model this situation.

Solution to 6.3.1: In this model, our starting value is 10 eagles, so let C be 10. Also, every two months there are 150% as many eagles as there were before, so a is 1.50 in the model. However, this time d is 2, as our value for a is based on values two months apart. Therefore, this gives us the following model:

$$y = 10 * (1.5)^{(x/2)}$$

©Brian Gillispie, 2016.

Example 6.3.2: 500 mg of a drug is given to a patient, and the amount of drug present is decreasing by 40% per 5 hours. Create an exponential model.

Solution to 6.3.2: In this model, 500 is our starting value, or C, in this model. This time we are given a percent decrease, so we have to convert it to a percent of, which, if we do, we find that this means that every 5 hours, the number of mg present is 60% of the number of mg present 5 hours ago. Therefore, this means that a in our model is 0.6, and d is 5, since the percent is based on figures from 5 hours ago. Plugging all of that into the model gives us:

$$y = 500 * (0.6)^{(x/5)}$$

This model has several advantages of the model used in Section 6.2, in that it is easier to read and use. Most modelers use this version of the model instead of the model given in Section 6.2 for that very reason. Plus, it can be used to solve the same type of questions we solved in Section 6.2 Let's demonstrate with a few Examples now.

Example 6.3.3: A bacteria colony has 250 cells, and the number of cells is increasing by 60% every 4 months. How many cells will be present after 20 hours pass?

Solution to 6.3.3: In this model, 250 is our starting value, so 250 is C in our model. Since we know that the number of cells is increasing every 4 hours by 60%, this means that every 4 hours the number of cells is 160% of the number of cells present 4 hours ago. Therefore, a is 1.6 in our model, and d is 4 hours. Plugging this into the model gives us:

$$y = 250 * 1.6^{x/4}$$

Now that the model is created, we can use it to predict how many cells will be present after 20 hours pass. Plugging in 20 for x gives us:

$$y = 250 * 1.6^{(20/4)}$$

©Brian Gillispie, 2016.

Which, calculates to give us:

$$y = 2621.44$$

So in 20 hours there will be about 2,621 cells present, rounded to the nearest cell.

Example 6.3.4: Using the same model as in Example 6.3.3, predict how long until there are 6,000 cells of bacteria present.

Solution to 6.3.4: Recall in Example 6.3.3, the model for the situation was:

$$y = 250 * 1.6^{x/4}$$

Since this time, we wish to know how long until there are 6000 cells present, we need to solve the following equation for x:

$$6000 = 250 * 1.6^{x/4}$$

Start by dividing both sides by 250:

$$24 = 1.6^{x/4}$$

Now, take natural log of both sides. This allows us to rewrite the side with the exponent to the following:

$$\ln(24) = \frac{x}{4}\ln(1.6)$$

Finish by dividing both sides by $\ln(1.6)$, and multiplying both sides by 4. Once that is done, we end up with:

$$x = \frac{4 * \ln(24)}{\ln(1.6)} \approx 27.05$$

Therefore, there will be 6000 cells present in about 27 hours.

Just like with the model in Section 6.2, we still have to remember that our starting value may require us to rescale what we plug into the model. Let's demonstrate that now.

Example 6.3.5: Due to inflation, an item that costs $25 in the year 2020 finds that every 7 years the price of the item is increasing by 70%. How much will the item cost in the year 2034?

Solution to 6.3.5: In this model, $25 is our starting value, so C is 25. Also we know that every 7 years the value of the item is increasing by 70%, so if we convert that to the appropriate percent of, we get that every 7 years the value of the item is 170% of the value of the item 7 years ago. Therefore, this means that in our model a will be 1.7, and d will be 7. Plugging all of this into our model gives us:

$$y = 25 * 1.7^{x/7}$$

To figure out how much this item will cost in the year 2034, we need to figure out how much time has elapsed since the model started on 2020. Turns out that 2034 is 14 years after 2020, so, we plug 14 into the model, which gives us:

$$y = 25 * 0.3^{(14/7)} \approx \$72.25$$

Therefore, in the year 2034, this item will now cost $72.25.

©Brian Gillispie, 2016.

6.3 Exercises

In Exercises 1 - 8, create an exponential model for each of the following situations.

1. In the month of January there are 100 rabbits, and the number of rabbits every 4 months is 255% of the number of rabbits that were present 4 months ago.

2. On the first week of school there are 25 students in the library on Thursday night, and estimates say that every month there will be 200% as many students as there were the previous month on Thursday nights.

3. A culture of bacteria has 100 cells, and the number of cells is increasing by 50% every 6 hours.

4. 1000 mg of a drug is given to a patient, and the amount of the drug present in the patient is decreasing by 33% per day.

5. A culture of bacteria has 500 cells, and the number of cells is increasing by 85% per 6 hours.

6. A store currently makes $5,000 in sales per day, and each month the amount of sales is increasing by 3%.

7. A radioactive material has 50 grams, and the number of grams present is decreasing by 35% every 20 hours.

8. A lake has 1000 catfish, and the number of catfish in the lake is increasing by 33% per 4 months.

©Brian Gillispie, 2016.

For Exercises 9 onward, create the model, and answer the question asked.

9. There are currently 10,000 people in a town, and every ten years, the number of people in the town is 110% of the number of people in the town 10 years ago. How many people will be present in 50 years, assuming an exponential model?

10. Population is booming due to an oil rush! Currently there are 882 people in town, and each month the number of people in the town is 130% of the number of people presently in the town. Assuming a 30 day month, how many people will be present in the town after 75 days, assuming an exponential model?

11. A medicine introduces 500 mg of a product into the human body, and the amount of the drug is decreasing by 43% per 8 hours. How much of the drug is present 14 hours from after the drug is taken, assuming an exponential model?

12. On an island, there are 100 rabbits, and the number of rabbits is increasing by 300% per 6 months. How many rabbits will be present in 44 months, assuming an exponential model?

13. Hyperinflation occurs occasionally in countries. In Germany, after World War I, the hyperinflation was so severe that an item that cost 41 Marks at the start of June 1921 increased by 1577% per 6 months! Assuming this trend continued, how much would this same item cost at the start of June 1922, assuming an exponential model?

14. A medicine introduces 100 g of a product into the human body, and the amount of the medicine is decreasing by 62% every 5 hours. How long until there is 30 g present, assuming an exponential model?

©Brian Gillispie, 2016.

15. In the year 2000, there were 15,000 people in a town, and the number of people in the town is increasing by 10% every 10 years. The state has promised to build a new arena in town once the population is 50,000 or more. In what year will the arena be built, assuming an exponential model?

16. In an effort to study for an exam overnight, you drink 500 mg of caffeine at 9 pm, and the amount of caffeine in your body is decreasing by 50% every 4.9 hours. How long until you only have about 30 mg of Caffeine present in your body, assuming an exponential model?

17. A lake has become polluted! If 25 liters of pollutant entered the lake on September 10th, and clean-up and filtering techniques are reducing the amount of pollutant by 75% per day, how much pollutant will be in the lake by September 13th?

18. You wish to buy a new phone, but currently the phone is too expensive at $1000. If the price of the phone is decreased by 25% per two months, and you wish to buy the phone when it costs $200, how many months will you have to wait to buy the phone?

©Brian Gillispie, 2016.

6.4: Half Life and Doubling Time Models

One situation where exponential models are commonly used is to model doubling time and half-life problems. In this Section, we will discuss what the words half-life and doubling time mean, as well as how to solve problems using those terms.

Let's start out by defining what the words half-life and doubling time mean. The word half-life means how long it takes for something to be reduced to half of the current amount. For instance, if we had 50 grams of something, and in 20 days we now had 25 grams, we would say the half-life of this object was 20 days. In other words, half-life can be thought of as how long it takes for something to decrease by 50%, or to have 50% of its' original amount left.

Similarly, the word doubling time means how long it takes something to double. For instance, if onions cost $2 today, and $4 45 days from now, we would say the doubling time of the price of onions is 45 days. In other words, doubling time can be thought of as how long it takes for something to increase by 100%, or to have 200% of its' original amount.

We summarize these two definitions below:

Definition 6.4.1: The **doubling time** is the time it takes for an item to have 200% of its original value, or, to increase by 100%.

Definition 6.4.2: The **half-life** is the time it takes for an item to decrease to 50% of its original value, or, to have 50% of its' starting value left.

While it is true these definitions could be used for non-exponential models, typically we only talk about doubling time and half-life for exponential models. Therefore, whenever you see the words doubling time or half-life, it is safe to assume an exponential model is to be used, unless you have a reason to not use such a model.

Now that we have half-life and doubling time defined, let's see how we can create a half-life or doubling time model. Recall the following model from Section 6.3:

$$y = C * a^{x/d}$$

©Brian Gillispie, 2016.

In Section 6.3, we defined this model for situations where we knew the percent the new value was of the old value, when it took d time intervals. Therefore, if we know the doubling time of something is time d, then, since doubling time means after d time steps the new value is 200% of the old value, we end up with the following for our model:

$$y = C * 2^{x/d}$$

Similarly, since half-life is the time it takes for something to have 50% of its current value. So, if something has a half-life of d, then the model for this situation is:

$$y = C * (0.5)^{x/d}$$

Let's summarize these results now:

Definition 6.4.3: If the starting value of an item is C, and the item has a doubling time of d, then the model is:

$$y = C * 2^{x/d}$$

However, if the starting value is C, and the item has a half-life of d, then the model is:

$$y = C * (0.5)^{x/d}$$

Be sure to use the correct model listed above for the situation. If the problem gives the half-life, then do not use the model for doubling time, and vice-versa. Now, let's work a few Examples to demonstrate these models.

6.4: Half Life and Doubling Time Models

Example 6.4.1: A culture of bacteria is growing rapidly. There are currently 100 cells, and the doubling time of the bacteria is estimated to be 4 hours. Create an exponential model for this situation.

Solution to 6.4.1: In this problem, we start with 100 cells, so our starting value (C) is 100. Our doubling time (d) is 4 hours, so plugging that into the model for doubling time gives us the following model:

$$y = 100 * 2^{x/4}$$

Example 6.4.2: A radioactive material has 10 grams present, and the half-life of the material is 55 days. Create an exponential model for this situation.

Solution to 6.4.2: In this problem, our starting value (C) is 10 grams, and our half-life (d) is 55 days. Plugging this into the half-life model gives us the following model:

$$y = 10 * (0.5)^{x/55}$$

Now, we will demonstrate using these models to make a few predictions.

Example 6.4.3: Carbon-15 has a half-life of 2.449 seconds, and there is currently 1000 grams of Carbon 15 present. How long until there is only 10 grams of Carbon-15 present?

Solution to 6.4.3: In this problem our starting value (C) is 1000 grams, and our half-life (d) is 2.449 seconds. Plugging all of that into the half-life model gives us the following model:

$$y = 1000 * (0.5)^{x/2.449}$$

Next, we wish to learn when there will be only 10 grams present. Plugging in 10 for y gives us:

©Brian Gillispie, 2016.

$$10 = 1000 * (0.5)^{x/2.449}$$

Dividing both sides by 1000 yields:

$$0.01 = (0.5)^{x/2.449}$$

Taking natural log of both sides, then using the rule of logs to bring the power down gives us:

$$\ln(0.01) = \frac{x}{2.449} * \ln(0.5)$$

Finally, we solve for x. Multiply both sides by 2.449, and divide both sides by $\ln(0.5)$ to end up with:

$$x = \frac{2,449 * \ln(0.01)}{\ln(0.5)} \approx 16.27$$

Therefore, in 16.27 seconds, you will have only 10 grams of Carbon-15 left.

Example 6.4.4: Under ideal conditions, the bacteria Escherichia coli have a doubling time of 17 minutes. If a culture of 50 cells of escherichia coli is kept in perfect laboratory conditions so they can double at this rate, how long until there are 400 cells of escherichia coli?

Solution to 6.4.4: First, we need to create our model. In this problem, our starting value (C) is 50 cells, and our doubling time (d) is 17 minutes. Plugging those into our doubling time model gives us the following:

$$y = 50 * 2^{x/17}$$

To find when the model will yield 400 cells, we plug 400 in for y, giving us:

©Brian Gillispie, 2016.

$$400 = 50 * 2^{x/17}$$

Divide by 50 to get:

$$8 = 2^{x/17}$$

Take the natural log of both sides, then use the rule of logarithms to bring the power down to end up with:

$$\ln(8) = \frac{x}{17} * \ln(2)$$

Multiply both sides by 17 and divide both sides by $\ln(2)$ to give us that the final answer for x is:

$$x = \frac{17 * \ln(8)}{\ln(2)} \approx 51 \text{ minutes}$$

A common question that arises in biology and chemistry is how long is the doubling time of something, based on how fast it is increasing. In order to solve these types of problems, first convert the percent increase to a percent of, then use the model from Section 6.3 to find how long until you have twice as much of the substance available. Similar reasoning will work for half-life problems as well. We will now demonstrate this with a couple of Examples.

Example 6.4.5: You currently have $25,000 in an investment account, and the amount of money present in the account is increasing by 5.5% per year. How long until your investment doubles?

Solution to 6.4.5: First, begin by creating the model. We know that our starting value (C) is 25,000. Also, we know that the money is increasing by 5.5% per year, so first we convert this to a percent of, which is 105.5% per year. That will make a be 1.055 in our model, and d will be 1 in our model, since the increase is per year. Plugging all of that into our model gives us:

©Brian Gillispie, 2016.

$$y = 25000 * (1.055)^x$$

Now, we wish to solve for when our investment will double. Double of 25000 is 50000, so we need to solve for when y is 50000. Plugging that in for y gives us:

$$50000 = 25000 * 1.055^x$$

Divide both sides by 25000 to get:

$$2 = 1.055^x$$

Next, take the natural log of both sides, then use the rule of natural logs to bring the power down. Doing that gives us:

$$\ln(2) = x * \ln(1.055)$$

Finally, divide by ln(1.05) to end up with:

$$x = \frac{\ln(2)}{\ln(1.055)} \approx 12.95$$

Therefore, it will take about 12.95 years for this investment to double.

Example 6.4.6: You have found a new radioactive substance, and wish to learn the half-life of this substance. All you know so far is that you started with 200 grams of the substance, and after 3 minutes the amount of substance had decreased by 10%. With that information, find the half-life of the substance.

Solution to 6.4.6: Start out by creating the model. You know that you started with 200 grams, so let that be your starting value (C). Also, since this substance is decreasing by 10% per 3 minutes, that means that every 3 minutes there is 90% of the amount that was present 3 minutes ago. By definition, this means a must be 0.90 and d must be 3 in the model. Plugging all of that into the model gives us:

$$y = 200 * (0.90)^{x/3}$$

Now, we wish to learn the half-life, so we need to solve for when we have half our starting value. Since we started with 200 grams, half of that is 100 grams. Plug that in for y gives us the following:

$$100 = 200 * (0.90)^{x/3}$$

Divide both sides by 200 to end up with:

$$0.5 = (0.90)^{x/3}$$

Take natural log of both sides, then use the rule of logs to bring the power down. This gives you:

$$\ln(0.5) = \frac{x}{3} * \ln(0.90)$$

Multiply both sides by 3, and divide both sides by $\ln(0.90)$ to end up with:

$$x = \frac{3 * \ln(0.5)}{\ln(0.90)} \approx 19.74$$

Therefore, the half-life of your new radioactive substance is 19.74 minutes.

©Brian Gillispie, 2016.

6.4 Exercises

For Exercises 1 - 8, create the exponential model for the situation.

1. A culture of bacteria has 150 cells, and doubles every 8 hours

2. A herd of deer has 25 deer, and doubles every 15.6 years.

3. A pond has 750 cells of algae, and the algae doubles every 5.23 days

4. A pond is polluted. Currently there is 10 g of pollutant in the pond, and the amount of the pollutant is doubling every 27.3 days

5. On a deserted island, the snake population is increasing. There are currently 91 snakes, and the population is doubling every 38.75 days

6. 15 kg of Cesium 137 is released into the atmosphere, and it has a half-life of 30.17 years.

7. There is 10 g of Titanium 44 present in a lab, and it has a half-life of 63 years.

8. You need Plutonium-238 to power your car so you can travel in time! You currently have 25 mg of Plutonium-238 and it has a half-life of 87.7 years.

6.4: Half Life and Doubling Time Models

For the remaining Exercises, create the exponential model, and answer the question asked.

9. You deposit $10,000 into a bank account that promises to double your money every 14 years. How long until you have $50,000?

10. Rabbits have invaded your farmland! There are currently only 10 rabbits, but they are doubling every week! How long until there are 100,000 rabbits on your farm?

11. Argon-39 has a half-life of 269 years. If you currently have 50 grams of Argon-39 present, how long until you have only 30 grams of Argon-39 left?

12. A population of salmon has 13000 salmon, and the salmon are doubling every 2.5 years. You will need to harvest the salmon when there are 20000 salmon present, else there will be too many salmon for the fish to survive. How long before you will need to harvest the salmon?

13. A patient takes a drug with 300 mg of a substance, and the amount of the substance present is decreasing by 2% per minute. What is the half-life of this substance?

14. According to the UN Website, the population of Australia has increased by 1.57% from the years 2010 to 2015. If the population of Australia is currently estimated to be 24 million, how long will it take for the population of Australia to double, based on this information?

15. You have discovered a new strain of bacteria. There is currently 100 mg of the bacteria present, and it is increasing by 50% every 4 hours. Assuming exponential growth, what is the doubling time of these bacteria?

16. You have discovered a new radioactive substance that looks really promising for scanning the human body! However, you do not know the half-life yet. Currently, there is 100 mg of the substance, and after 4 minutes the amount of the substance has decreased by 10%. Assuming an exponential model, what is the half-life of this new substance?

©Brian Gillispie, 2016.

17. You have invested $50,000 into an account which promises to increase your investment by 3.32% per year. What is the doubling time of this investment?

18. You decide at the last minute to only invest $1,000 into that account which promises to increase your investment by 3.32% per year. What is the doubling time of your investment now?

19. Actually, you decided to invest a lot of money, $250,000 into that account which promises to increase your investment by 3.32% per year. What is the doubling time of your investment now?

20. Based on your answers to Exercises 19 - 21, what observation can you make about how the starting value affects your doubling time?

21. A pollutant has escaped into a pond. If currently there is 100 liters of the substance, but it is decreasing by 17% per day, what is the half-life of the pollutant?

22. A flask of Hydrogen-3 has 2 liters of gas, and is decomposing at the rate of 0.005% per month. What is the half-life of Hydrogen-3, in months?

23. A population of owls has 150 owls, and the number of owls is increasing by 10% per year. What is the doubling time of the population?

24*. You invest $150,000 into an account that is increasing your investment by 3.56% per year. How long until your investment triples?

©Brian Gillispie, 2016.

6.5: Exponential Models from Two Data Points

In the previous Sections, we have covered how to create exponential models when we are given either information about how much the starting value is supposed to increase per time value, or how much the new value is supposed to be of the starting value after a certain amount of time passes. However, in reality, that information is usually not known. Instead, we usually have to figure out that information in order to create our exponential models, which we can do as long as we have two or more data points. Let's demonstrate how this can be done.

Suppose you have been monitoring the decay of a radioactive substance over time. You started with 100 grams of the substance, and after 4 hours have passed, you have 65 grams of the stuff. Now, in order to use an exponential model, we need to figure out how much the new value, the one from 4 hours later, is of the old starting value. By the definition of percent of, that value is calculated by:

$$\frac{65}{100} = 0.65$$

$y = C \cdot q^{(x/d)}$

This means that after 4 hours have passed, the new value is 65% of the old value. We can then use our model from Section 6.3, which will give us the following once we plug in the appropriate numbers:

$$y = 100 * (0.65)^{x/4}$$

This means that as long as we have two points, we can find what the percent the new value is of the old, which will then allow us to use the exponential model from Section 6.3.

Let's demonstrate this now with a few more Examples before proceeding.

©Brian Gillispie, 2016.

Example 6.5.1: A group of bunnies starts out with 100 bunnies, and 6 months later there are 250 bunnies. Create an exponential model for this situation.

Solution to 6.5.1: In this problem, we know that we started with 100 bunnies, and 6 months later we had 250 bunnies. We start by computing the percent the new value is of the old, which is calculated by:

$$\frac{250}{100} = 2.50$$

This means that after 6 months have passed, we have 250% of the initial 100 bunnies. This means that a in our model will then be 2.5, and d will be 6, as the percent of is based on a figure 6 months later. Also, C will be 100, as that is our starting value. Plugging all of this into our model gives us:

$$y = 100 * (2.5)^{x/6}$$

Example 6.5.2: A radioactive explosion has taken place! When the explosion occurs there was 10 g of a radioactive substance, and 5 years later there is still 8.5 g of the radioactive substance. Create an exponential model for this situation.

Solution to 6.5.2: In this problem, we know that we started with 10 g, and after 5 years we have 8.5 g. To compute the percent the new value is of the old, we calculate:

$$\frac{8.5}{10} = 0.85$$

This means that after 5 years have passed, 85% of the substance remains. This means that a in our model will then be 0.85, and d will be 5, as the percent of is based on a figure 5 years later. Also, C will be 10, as that is our starting value. This means that our final exponential model is the following:

©Brian Gillispie, 2016.

$$y = 10 * (0.85)^{x/5}$$

Once we have the model, we can use them to make predictions, like we did in the previous Sections. Let's demonstrate.

Example 6.5.3: Initially there are 125 bats in a cave, and after 6 months there are 188 bats in the cave. Using an exponential model, predict how many bats will be in the cave after 9 months.

Solution to 6.5.3: In this situation, we know we have 125 bats, and after 6 months there are 188 bats. To compute the percent the new value is of the old, we compute:

$$\frac{188}{125} = 1.504$$

This means that after 6 months have passed, the number of bats present is 150.4% of the previous number of bats. Therefore, a is 150.4 in our model, and d is 6. Since we started with 125 bats, C is 125. Therefore, our exponential model is:

$$y = 125 * (1.504)^{x/6}$$

Next, we wish to know how many bats are present after 9 months have passed. Plug in 9 for x to get:

$$y = 125 * 1.504^{(9/6)} \approx 230.56$$

Which means that, after 9 months have passed, we predict there will be 231 bats, rounded to the nearest bat.

Watch out when creating these models, as the way our exponential model is designed, the value for C must always be the value you want the model to return when x is 0. This will force you to scale x accordingly. Let's demonstrate.

©Brian Gillispie, 2016.

Example 6.5.4: Tuition at a college is $4,500 a semester in 2010, and $4,600 a semester in 2011. Using an exponential model, predict how much tuition will be in the year 2017.

Solution to 6.5.4: In this situation, we are given two tuition values, and they are one year apart. This means that we can calculate the percent the new value is of the old, by calculating:

$$\frac{4600}{4500} \approx 1.0222$$

Which means that the tuition in 2011 is approximately 102.22% of the tuition in 2010. This means then that a is 1.0222 in our model. To find d, we need to know how much time passed before our second tuition value, which in this case is found by computing $2011 - 2010$, which is 1. Therefore, d is 1 in this model. Finally, our starting tuition value is 4500, so that will be C. Plugging all of this into our model gives us the following:

$$y = 4500 * (1.0222)^x$$

Now, we wish to learn what tuition is in the year 2017. Be careful though, as the way we set up the model, an input of 0 returns 4500, so that means x is in years after 2010! Therefore, to find the tuition in the year 2017, we have to plug in 7, which, if we do that, we get:

$$y = 4500 * (1.0222)^7 \approx \$5,247.64$$

Therefore, this model predicts that in the year 2017, tuition will be about $5,247.64 per semester.

Notice in the last Example we had to calculate d for the model, as it was not directly given. In general, if we know two points, we can always compute d by calculating $x_2 - x_1$.

So far in all of our Examples the starting value has been given, but sometimes, due to scaling, that is not possible. Let's demonstrate with our next Example.

©Brian Gillispie, 2016.

Example 6.5.5: Your boss wants the model in Example 6.5.4 set up so that x means years after 2000, instead of years after 2010. Create an exponential model for this situation.

Solution to 6.5.5: This time it is a little more complicated, as you need to shift your x value. Thankfully, we can easily shift our solution to Example 6.5.4 to meet this demand. If x is to be in terms of years after 2000, then an input of 10 should yield $4500, and an input of 11 should yield $4600, as those inputs now stand for 2010 and 2011, respectively. Therefore, we can easily shift the time scale by doing the following:

$$y = 4500 * (1.02222)^{x-10}$$

This method in Example 6.5.5 works, and is important to remember, because sometimes we do not know the starting value for our model! When that happens, treat the value that is earliest in time as the value for C, then once the model is done, offset the x value by as much as needed.

Now that we have seen via Examples all of the different factors to keep in mind when working with two points, let's summarize all that we have learned below:

Definition 6.5.1: If we know two data points (x_1, y_1) and (x_2, y_2), then an exponential model of the form

$$y = C * (a)^{(x-z)/d}$$

Can be found, with C, a, d, and z defined as follows:

$$C = y_1$$
$$a = \frac{y_2}{y_1}$$
$$d = x_2 - x_1$$
$$z = x_1$$

©Brian Gillispie, 2016.

Notice that C is no longer the starting value in our model with these changes. In general, C will **not** stand for the starting value in an exponential model, but will stand for a constant that we have to solve for. So far, we have not had to worry about that, but going forward it will be possible.

Let's now demonstrate this new model on solving a problem where we do not know the starting value, but we have two points.

Example 6.5.6: On the fifth month of observation, there are 30 ducks on a pond, and on the ninth month of observation there are 50 ducks on a pond. How many ducks were on the pond during the third month of observation?

Solution to 6.5.6: In this example, we do not know the starting value. However, we know that on month five there were 30 ducks, and on month nine there were 50 ducks. Therefore, our two points are $(4, 30)$ and $(9, 50)$, with x in months and y in ducks. Using the rules from Definition 6.5.1, we first calculate a as:

$$a = \frac{50}{30} = 1.6667$$

Next, we find d, which is the difference between the x values. In this case, d is $9 - 4$, or 5. C is our y_1, which is 30, and z is our x_1, which is 4. Putting all of this together, we end up with the following for our model:

$$y = 30 * (1.66667)^{(x-5)/4}$$

Now, we answer the question asked. We want to know the number of ducks back on month three. Since x is in months, we plug 3 into the equation for x, and end up with:

$$y = 30 * (1.6667)^{(3-5)/4} \approx 23.24$$

Therefore, we conclude that back on month three, there were about 23 ducks on the pond.

©Brian Gillispie, 2016.

6.5 Exercises

In Exercises 1 - 8, you are given two data points. Create an exponential model. For Exercises 7 and 8, be sure your model is set up to meet the restrictions given!

1. A herd of deer has 50 deer, and 8 months later there are 75 deer

2. A culture of bacteria has 150 cells, and 9 hours later it has 325 cells

3. A school of fish has 35 fish, and 10 months later it has 50 fish.

4. There are currently 10 grams of a radioactive material, and 9 years later there are 8.7 grams of the material.

5. A small town has 1,000 people, and 3 years later it has 900 people.

6. A lake has 25 grams of a pollutant, and 3 years later it has 24 grams of the pollutant.

7. A pond has 20 ducks after 4 years pass, and 60 ducks after 10 years pass. An input of zero should mean the year zero.

8. A school has 175 freshmen enrolled 25 years after the school opens, and the school has 200 freshmen enrolled 35 years after the school opens. An input of zero should mean the year the school opened.

Using the techniques of this Section, create the exponential model. Then, using the techniques of the previous Sections answer the question asked.

9. Your starting salary is $50,000 a year, and in 5 years your salary will be $65,000 a year. How much will you be making in 8 years at this rate?

10. A culture of bacteria has 100 cells, and after 4 hours it has 200 cells. How long until there are 300 cells?

11. A pack of wolves has 25 wolves, and 4 years later the pack has 30 wolves. How long until the pack has 35 wolves at this rate?

12. A lake has 10 grams of a pollutant, and 3 months later, the lake has 9 grams of the pollutant. How long until the lake has only 1 gram of the pollutant?

13. 53 mSV of a radioactive material[88] is used in a medical scan, and an hour later 50 mSV are still present in the body. How much of the radioactive material will still be present in the body 24 hours later?

14. A radioactive material has 50 grams after 20 years have passed, and 40 grams after 40 years have passed. Create an exponential model for this situation, and use the model to predict how many grams will be present in 80 years?

15. A researcher notices that there were 100,000 deer in the wild in the year 2010, and 120,000 deer in the wild in the year 2015. Create an exponential model and use the model to predict how many deer are present in the year 2020.

---

[88] 53 mSV stands for 53 microsieverts, a unit used to measure radiation doses [20].

©Brian Gillispie, 2016.

6.6: Application: Examples of Some Exponential Models

Throughout this Chapter we have shown various ways an exponential model can be created with the given data provided. At this point we have covered as much as is possible with the current version of the exponential model that we are using in this Chapter. However, there are other exponential models in existence that are commonly used, and those models are what we will discuss in this Section. We are not going to cover how to create these models from a data set, as that will require a background in differential equations. However, I do encourage the reader who is interested in taking these models further to consider studying differential equations at some point in time.

6.6.1: Continuous Growth Model

The models we have used in this Chapter are called discrete growth models, as they assume that the growth occurs at fixed points in time. If we assume that the growth is always occurring, then we are using what is called a continuous growth model. Creating the continuous growth model requires that we solve the following equation:

$$\frac{\partial y}{\partial t} = ky, y = C$$

This equation is known as a first order differential equation [13]. We are not going to cover how the equation is arrived at, or how to solve it here, but if you are interested in those details they are covered in most textbooks on differential equations. Instead, we are going to focus on the answer, which is the model we are going to look at. The solution to this differential equation is given by the following:

$$y = C * e^{kt}$$

©Brian Gillispie, 2016.

Incidentally, this is the model that most textbooks use for exponential modeling. In this model, C stands for the starting value, and k stands for the continuous growth factor. However, unlike our model, where the growth factor can be computed quickly, the continuous growth factor is often unknown and not easy to compute either. Still, this model is commonly used in half-life and doubling time equations, so it is worthwhile knowing. Let's demonstrate how to use this model now with an Example.

Example 6.6.1: A culture of bacteria has 100 cells, and it doubles every 4 hours. How long until there are 300 cells?

Solution to 6.6.1: To use our continuous growth model, we need to first figure out what our data points are. We have two points based on this data, and they are given by $(0, 100)$, and $(4, 200)$, with the second point coming from the fact that we know that the cells double every 4 hours, and double of 100 is 200.

Now we are going to use these two points to find our constants C and k by plugging the points into the continuous growth model equation. Plug the point $(0, 100)$ into our equation to get the following:

$$100 = C * e^{k0}$$

But, k times zero is zero, and e to the zero power is one. This gives us $C * 1$, which is C. Therefore, we conclude that C is 100 in this equation. Let's use this data to update our model now:

$$y = 100 * e^{kt}$$

Now we need to find a way to find k, which we will do by plugging into the model the other data point. Plugging in $(4, 200)$ gives us the following:

$$200 = 100 * e^{4k}$$

©Brian Gillispie, 2016.

Dividing both sides by 100 gives us:

$$2 = e^{4k}$$

This is an exponential equation, which we can solve by taking the natural log of both sides. If we do that, we can bring the power of 4k down out of the exponent, which gives us the following:

$$\ln(2) = 4k * \ln(e)$$

We need to solve this equation for k, and we do that by dividing both sides by 4 and $\ln(e)$. Doing this gives us the following value for k:

$$k = \frac{\ln(2)}{4 * \ln(e)} \approx 0.173287$$

However, to avoid rounding error, we will not use the decimal form for k here. However, we can simplify our answer by noticing that $\ln(e) = 1$, so that means our k is really $\frac{\ln(2)}{4}$. We will use this in our equation as follows:

$$y = 100 * e^{(\frac{\ln(2)}{4}*t)}$$

Now we can finally answer the question asked. We want to know when there are 300 cells present. Plugging in 300 for y gives us:

$$300 = 100 * e^{(\frac{\ln(2)}{4}t)}$$

We need to solve this equation for t to answer the desired question. To solve for t, we divide both sides by 100, then take natural log of both sides, giving us:

©Brian Gillispie, 2016.

$$\ln(3) = \frac{\ln(2)}{4} * t * \ln(e)$$

Solving for t, and remembering that $\ln(e) = 1$ we get the following final answer for t:

$$t = \frac{4 * \ln(3)}{\ln(2)} \approx 6.33985 \text{ hours}$$

Now, let's look at this Example for a minute. If we had used the model we developed back earlier in this Chapter, we would have ended up with the following model:

$$y = 100 * 2^{(x/4)}$$

Then, to find when we had 300 cells, we would have just had to plug in 300 for y, divide both sides by 100, take natural log of both sides and solve for x to get:

$$x = \frac{4 * \ln(3)}{\ln(2)}$$

This is the exact same answer as the new model, but with a lot less steps. This is why we did not cover this model in the book when we were creating models, as our model gets the exact same answer in a lot less steps. Plus, there is no practical way to measure continuous growth, as our eyes only see in discrete time steps! Therefore, any model with a data set is going to be better served with the models we stated, as it is easier to work with. This here is an example of the tradeoff between being precise, or sacrificing some precision for an easier to work with model. Our model in this chapter is not as mathematically precise, as our model is not the solution to the differential equation. Despite that, our model still gives reasonably accurate solutions, and is easier to work with than the more precise model $y = C * e^{kt}$.

Many other exponential models are based on a modified version of the model $y = C * e^{kt}$. So, understanding this model and the steps we took above is needed to understand these newer models. I encourage the reader to return to some of our Section 6.5 problems and try to

©Brian Gillispie, 2016.

create the model $y = C * e^{kt}$ for those problems now, as the practice will be invaluable when we discuss the next two models.

6.6.2: The Logistic Model

Another commonly used exponential model is the logistic model, which attempts to fix a major problem in the exponential model we used in this Chapter. To see what this problem is, consider the following model for the number of bunny rabbits in Australia, with x years after 1859:

$$y = 24 * (6.25)^x$$

Now, let's use this model to predict how many rabbits there were in Australia in the year 1959. Plugging in 100 into our equation gives us:

$$y = 24 * 6.25^{100} \approx 9.29 * 10^{80}$$

Obviously this is impossible an impossible number of rabbits, because before that many rabbits would exist they would have eaten all of the food available on the planet, and then starved themselves to death. The current exponential model makes no allowance for things like population caps and other details that would eventually slow down growing populations. The logistic model[89], on the other hand, adds the population cap into the model, and is as follows:

Definition 6.6.1: The **Logistic Model** is defined as follows:

$$y = \frac{M}{1 + C * e^{kt}}$$

Where:

$M = $ *The upper limit on the population*

$C, k = $ *constants that must be solved for*

Unlike with the previous models, C is no longer the starting value! This time C stands for a constant that we do not know, and have to solve for it. Likewise, we have to solve for k in a similar fashion.

The logistic function has the advantage that the function no longer increases without bound. The function will increase quickly still for small populations, but as the size of the population increases towards the population limit, then the growth of the population will slow down. Let's demonstrate this with an Example.

Example 6.6.2: Given the logistic model $y = \frac{1000}{1+499*e^{-0.20t}}$, with t in hours, calculate the value for each hour from 0 to 15.

Solution to 6.6.2: To calculate the value for each hour, we need to plug in the given value for t and calculate it. Doing that gives us the following:

Time	Output
0	2
1	2.44
2	2.98
3	3.64
4	4.44
5	5.42
6	6.61
7	8.06
8	9.83
9	11.98
10	14.59
11	17.76
12	21.61
13	26.27
14	31.90
15	38.69

In that Example, it appears the values are growing without bound. So, it at first appears the upper bound of 1000 is not doing anything. However, if we plug in $t = 50$, we get the output of 977.8. Plug in $t = 75$, we get an output of $y = 999.85$. Now the growth is slowing down, as there is hardly any growth between an input of 50 and 75. This makes sense practically too, as

©Brian Gillispie, 2016.

when our input was 50, our output was near the population limit. Therefore, we would expect growth to slow down, as the current population is now spending more time competing for resources instead of growing.

The major weakness of the logistic model is it is more complicated than the other models in this Chapter. Therefore, it is up to the modeler to decide if the extra accuracy is worth the tradeoff in complexity. In general, if one is only doing short term projections, and the population is believed to be nowhere near the cap, the models of Sections 6.2 and 6.3 work well. However, if one is working calculations that are near the population cap, then the logistic model should probably be used.

Let's now put all of this together and show how to create a logistic model.

Example 6.6.3: In 1859 Thomas Austin released 24 rabbits into Australia [11]. Estimates say there were 150 rabbits in 1860. If the population cap of the rabbits is estimated to be 900,000,000, create a logistic model for this situation.

Solution to 6.6.3: Start off by figuring out what we already know. We know that the population cap is 900,000,000, so that is M in our model. Also, we know two points, the number of rabbits in 1859 and the number of rabbits in 1860. Since it will be easier to solve for C and k if one of the inputs is 0, we will set t to be years after 1859. This then makes our two points $(0, 24)$ and $(1, 150)$ for this model.

Now, proceed by plugging in $(0, 24)$ and solving for what we can. If we do that, our equation is:

$$24 = \frac{900,000,000}{1 + C * e^{k(0)}}$$

Recall that $k * 0 = 0$, and then $e^0 = 1$. This gives us the equation:

$$24 = \frac{900,000,000}{1 + C}$$

We can solve this for C. To do this, flip both sides of the equation, which gives us:

$$\frac{1}{24} = \frac{1+C}{900{,}000{,}000}$$

Multiply both sides by 900,000,000 to get:

$$37{,}500{,}000 = 1 + C$$

Subtract 1 to get that $C = 37{,}499{,}999$. This allows us to update the logistic model, since we now know the value for C. If we do this, we get:

$$y = \frac{900{,}000{,}000}{1 + 37499999 * e^{kt}}$$

Now we need to figure out what k is. We already used the point $(0, 24)$ to find C, so let's try plugging in the point $(1, 150)$ this time. Plugging that into the equation gives us the following:

$$150 = \frac{900{,}000{,}000}{1 + 37499999 * e^{k(1)}}$$

Since the variable we want to solve for is in the denominator of a fraction, flip both sides of the equation. Doing that gives us:

$$\frac{1}{150} = \frac{1 + 37499999 * e^{1k}}{900{,}000{,}000}$$

Now we need to isolate the exponential term. To do that, multiply both sides by 900,000,000. This gives us:

$$6000000 = 1 + 37499999 * e^{1k}$$

©Brian Gillispie, 2016.

Subtract the 1, then divide both sides by 37,499,999 to get the exponential term isolated, as follows:

$$0.15999998 = e^{1k}$$

Take natural log of both sides to get:

$$\ln(.15999998) = 1 * k * \ln(e)$$

Recall that 1 times k is k, and $\ln(e) = 1$. Making those substitutions gives us:

$$k = \ln(.15999997) \approx -1.832581651$$

Plugging this into our model gives us our final logistical model of:

$$y = \frac{800,000,000}{1 + 33,333,332.\overline{3}e^{-1.832581651t}}$$

Incidentally, if we now use this model to predict the number of rabbits present in 1959, we would get that there were only 900,000,000 rabbits, or the population limit. That's a much better (and more realistic) prediction than a prediction claiming that in 1959 Australia had so many rabbits that one could not walk in Australia without stepping on a rabbit!

The logistic model still has a few problems present. For one, once the population reaches the population limit it can never leave. So in our rabbit model, once there are 900,000,000 rabbits, there will always be 900.000,000 rabbits, no more and no less. Obviously this is not completely realistic, as some rabbits will die and more will enter the population. So at the minimum there will be a brief period of 899,999,999 rabbits and a brief period of 900,000,001 rabbits. However, the model does not capture this. More importantly though, the model makes no allowance for outside factors that may affect and change the growth of the rabbit population. However, accounting for more outside factors would complicate the model more, so in practice most modelers do not add those details into the equation.

©Brian Gillispie, 2016.

6.6.3: Newton's Law of Cooling

Newton's Law of Cooling is another exponential model that is commonly used. The model is based on the observation that the rate an object cools is proportional to the difference in temperature between the object and the surrounding medium. You have probably noticed Newton's Law of Cooling in action before. The minute you remove a turkey from the oven the inside temperature of the turkey is usually 165 degrees or so. However, if you let the turkey sit for a few hours, it will eventually cool down to the temperature of the room. All of this is reflected in the following model for for Newton's Law of Cooling[90], as follows:

Definition 6.6.2: The model for **Newton's Law of Cooling** is as follows:

$$T = T_m + (T_0 - T_m)e^{kt}$$

Where:

$T = $ *Temperature of the object t hours later*

$T_m = $ *Temperature of surrouding medium*

$T_o = $ *Temperature of the object at time zero*

Let's now demonstrate the use of this model in a few Examples.

---

[90] Some people define the model with the k term negative. Doing that means the signs will cancel out when solving the equation for k, so the sign is not relevant as k will always pick up the appropriate sign via the math. To make the math easier we chose to omit the minus sign for that reason.

©Brian Gillispie, 2016.

Example 6.6.4: A police detective is called to investigate a crime scene, where a body is found. The body is found by the detective at 12:39 am, and the body temperature is currently 92 degrees. Two minutes later, the body temperature is taken again, and it is now 91.7 degrees. If the temperature of the room where the body was found is currently 68 degrees, what time did the murder occur?

Solution to 6.6.4: To solve this we are going to have to decide on what time to set our zero at. Initially it is tempting to set the zero at the time the murder occurred, but we don't know when that was! Therefore, we will instead set our zero at the time the first temperature was taken. Therefore, in this problem, a time of zero means 12:39 am, and t will be in minutes after 12:39 am. That makes our data points then $(0, 92)$ and $(2, 91.7)$.

Next, figure out what are the values for T_m and T_0 in the model. Recall that T_0 stands for temperature at time zero, and we set our time zero to be 12:39 am. This means that our T_0 needs to be 92 then, as that is the temperature at 12:39 am. T_m stands for the temperature of the surrounding medium, and in this case that is the air temperature of the room, which is 68 degrees. Therefore, our T_m is 68.

Plug all of this into the Newton's Law of Cooling Model to get the following for our model:

$$T = 68 + (92 - 68)e^{kt}$$

Now, we will plug in our data points to find k. Plugging in $(0, 92)$ will go nowhere though, as $0 * k = 0$, and we need to solve for k. So we will instead plug in $(2, 91.7)$. Doing that, and noting that $92 - 68 = 24$ gives us the following:

$$91.7 = 68 + 24 * e^{2k}$$

Subtract 68 from both sides to get:

$$23.6 = 24 * e^{2k}$$

Divide both sides by 24 to get:

$$\frac{23.6}{24} = e^{2k}$$

Take natural log of both sides and use the rule of logarithms to get:

$$\ln\left(\frac{23.6}{24}\right) = 2k * \ln(e)$$

Remembering that $\ln(e) = 1$ and dividing both sides by 2 gives us the following for k:

$$k = \frac{\ln(\frac{23.6}{24})}{2} \approx -0.0242375$$

With k solved for, our model is now the following:

$$T = 68 + 24 * e^{-0.0242375t}$$

Now we need to solve for when the murder occurred. To do this we need to make an assumption, and that is we will assume that the person murdered was not sick, and therefore had a body temperature of 98.6 degrees when murdered. So, to find the time of the murder we need to solve the equation and find when the person was 98.6 degrees. Plug in 98.6 for T, and solve for t to finish the problem now.

$$98.6 = 68 + 24e^{-0.0242375t}$$

Subtract 68 from both sides to get:

$$30.6 = 24 * e^{-0.0242375t}$$

©Brian Gillispie, 2016.

Divide both sides by 24 now:

$$1.275 = e^{-0.0242375t}$$

Take natural log of both sides, and use the rule of logarithms to get:

$$\ln(1.275) = -0.0242375t * \ln(e)$$

Finally, using the fact that $\ln(e) = 1$ and dividing both sides by -0.0242375 gives us the following for t:

$$t = \frac{\ln(1.275)}{-0.0242375} \approx -10.02 \text{ minutes}$$

This means that the murder occurred about 10.02 minutes before the time we declared time zero. Since time zero was 12:39 am, the detective can conclude that the murder occurred sometime between 12:28 and 12:29 am.

©Brian Gillispie, 2016.

6.6 Exercises

Problems with an E in front of them require the Exponential equation as covered in Section 6.6.1. Problems with an L in front of them require the Logistic model as covered in Section 6.6.2. Problems with an N in front of them require the Newton's Law of Cooling model as covered in Section 6.6.3.

E1: There are 100 rabbits currently in the wild, and it is believed that in 2 years there will be 150 rabbits in the wild. Using the continuous exponential model, calculate how many rabbits will be present in the wild in 4 years.

E2: A culture of bacteria has 100 cells, and 4 hours later it has 200 cells. Using the continuous exponential model, predict how many cells will be present after 8 hours have passed.

E3. A radioactive material has 50 grams of the material present, and after one year there are 49.5 grams of the substance present. Using the continuous exponential model, predict how long until only 25 grams of the substance are present.

E4: Repeat E3, but this time use the half-life model we used in Section 6.4. What is the difference between the answer you found in Exercise E3 and in Exercise E4?

L1: A population of wolves is released into the wild. Initially there are 50 wolves released, then one year later a spotter counts 80 wolves in the wild. If the area the wolves were released into is estimated to be able to support 1,000 wolves, how many wolves will there be in 5 years after the wolves were released?

L2: Using the model created in Exercise L1, calculate how long until there are 500 wolves?

L3: The United States had 249.6 million people in 1990, and 282.2 million people in 2000. Assume that the United States can only support a population of 600 million people, and then predict the population of the United States in 2020 with your model.

©Brian Gillispie, 2016.

L4: Repeat Exercise L3, but this time assume the United States can support one billion people.

L5: A colony of bacteria is growing rapidly in a tube. If there are 100 cells at 1 pm and 125 cells at 1:05 pm, and if the most the tube can support is 1000 cells of bacteria, calculate how long until there are 800 cells in the tube.

N1: A turkey is removed from an oven. When the turkey is removed, the internal temperature of the turkey is currently 170 degrees, and 5 minutes later the turkey is currently 141 degrees. Assuming the room temperature is 72 degrees, how long until the temperature of the turkey is 115 degrees?

N2: You have a cup of coffee that is currently 150 degrees, and after 2 minutes have passed, the coffee is now 130 degrees. If the room temperature is 70 degrees, how long until the cup of coffee is 90 degrees?

N3: A detective was called to a hotel as a body was found in Room 412. If the body is 88 degrees when the detective gets there at 1:15 am, and 87.8 degrees at 1:30 am, if the room temperature is 72 degrees, at what time was the crime committed? *Hint: Remember to assume the body was 98.6 degrees at the time the crime occurred.*

N4: A frozen steak is placed into an oven. When the steak is placed in the oven, the steak is currently 5 degrees, and after being in the oven for 5 minutes, the steak is now 32 degrees. If the oven is set at 400 degrees, how long until the steak is 165 degrees?

©Brian Gillispie, 2016.

Chapter 6 Summary

6.1: In this Section, we reviewed what exponents and logarithms are. In addition, we discussed how to solve exponential equations by using logarithms.

6.2: in this Section we covered how to create an initial exponential model. If our starting value is given by C, and we know that the next value is a certain percent of the previous value, then we can model this (with a in decimal form) by the following:

$$y = C * a^x$$

If we are given a percent increase or a percent decrease instead, this model can still be used, provided the percent increase or decrease is first converted to a percent of.

6.3: In this Section we covered how to model an exponential model if the percent the next value is of the previous is not one time interval apart. This changed our model to the following, with d the timer interval the percent is in:

$$y = C * a^{x/d}$$

6.4: In this Section, we covered how to create models involving half-life and doubling time. If we have a stating value C, and a doubling time d, then the situation can be modeled by:

$$y = C * 2^{x/d}$$

If, instead, we know that our half-life is d, then the situation can be modeled by:

$$y = C * 0.5^{x/d}$$

©Brian Gillispie, 2016.

6.5: In this Section, we covered how to create an exponential model if two data points are provided. If we have two data points (x_1, y_1), (x_2, y_2), then the data points can be modeled by the following exponential model:

$$y = C * a^{(x-z)/d}$$

With C, a, d and z defined as follows:

$$C = y_1$$
$$a = \frac{y_2}{y_1}$$
$$d = x_2 - x_1$$
$$z = x_1$$

©Brian Gillispie, 2016.

6.6: In this Section, we discussed three other exponential models. The first model we looked at was:

$$y = Ce^{kx}$$

This model assumes continuous growth, and is the solution of the differential equation where population growth is proportional to the amount present.

The second model we discussed was the logistic model, which caps the population at M. The logistic model is given by:

$$y = \frac{M}{1 + Ce^{kt}}$$

In this model, C no longer stands for the starting value, and has to be solved for individually now.

The third model discussed in Section 6.6 was Newton's Law of Cooling, which is given by:

$$T = T_m + (T_0 - T_m)e^{kt}$$

Where:

$$T = Temperature\ of\ the\ object\ t\ hours\ later$$
$$T_m = Temperature\ of\ surrouding\ medium$$
$$T_o = Temperature\ of\ the\ object\ at\ time\ zero$$

©Brian Gillispie, 2016.

Chapter 7: Other Continuous Models

Now that we have explored linear and exponential models and how to use them, we will turn our attention to various other continuous models in this Chapter. We will begin by discussing how to use a quadratic to model the height of an object thrown into the air. From there we will move onto polynomial interpolation, then to proportionality, then to trig models, and finally we will discuss a little bit about log scaling data points and models, a technique often used when one of the coordinates has both really small and really large numbers.

©Brian Gillispie, 2016.

7.1: Some Quadratic Modeling

The first continuous model we will discuss in this Chapter is the quadratic model. In this Section, we will explore and discuss one of the more well-known quadratic models, the height equation.

7.1.1: The Height Equation

A quadratic model can be used to model the height of an object after it is tossed into the air, if we ignore air resistance[91]. The height of the object y (in feet) after t second has elapsed is given by the following:

$$y = -\frac{g}{2}t^2 + v_0 t + h_0$$

In this model, g stands for the acceleration due to gravity (usually 32 ft/sec^2), v_0 stands for the initial velocity of the object, and h_0 stands for the initial height.

Before we go any further, let's demonstrate setting up this model in a couple of examples.

Example 7.1.1: An object is tossed into the air with an initial velocity of 16 ft/sec from 3 feet in the air. Create the quadratic model if $g = 32$ ft/sec^2

Solution to 7.1.1: In this problem, our initial velocity (v_0) is 16, and our initial height (h_0) is 3 feet. Plugging those values into the equation with our given value of g gives us the following for our model:

$$y = -16t^2 + 16t + 3$$

---

[91] This model also assumes the object never reaches escape velocity.

©Brian Gillispie, 2016.

Example 7.1.2: A cannonball is fired from a height of 25 feet above the ground, with an initial velocity of 500 ft/sec. Create a quadratic model if $g = 32$ ft/sec^2.

Solution to 7.1.2: In this problem, our initial velocity (v_0) is 500, and our initial height (h_0) is 25. Plugging those values into the equation with our given value of g gives us the following for our model:

$$y = -16t^2 + 500t + 25$$

Now that we know how to create a quadratic model given the initial velocity, the initial height, and the acceleration due to gravity, we can now proceed to discuss how to use these models. Common uses of these models are to compute when an object hits the ground, or when does an object thrown into the air reach a desired height.

7.1.2: The Zeros of a Quadratic

If we wish to find when an object thrown into the air hits the ground, we need to be able to find the zeros of the polynomial, as the ground is represented by zero height in the model. This means we will need to be able to solve the following equation:

$$at^2 + bt + c = 0$$

This equation has a known solution, which is given by the following.

Definition 7.1.1: The zeros of a quadratic of the form $at^2 + bt + c$ can be found by the formula:

$$t = \frac{-b \pm \sqrt{b^2 - 4ac}}{2a}$$

©Brian Gillispie, 2016.

While it is true that we can sometimes use factoring, completing the square or the square root property to solve for the zeros of a quadratic, in practice real world data results in an equation that is not factorable into integers. Therefore, we will focus only on solving with the quadratic formula in this book, even though there are other methods that can and do work as well for solving this equation.

Let's now use the quadratic formula to demonstrate how to find the zeros of the quadratic in an Example before proceeding.

Example 7.1.3: An object tossed into the air is modeled by the quadratic equation $y = -16t^2 + 16t + 3$. When does the object hits the ground?

Solution to 7.1.3: For this problem, notice that for the quadratic formula, a corresponds to the coefficient on the t^2 term, b corresponds to the coefficient on the t term, and c corresponds to the constant term. That means for this equation, a is -16, b is 16, and c is 3. Plugging these numbers into the quadratic formula gives us the following:

$$t = \frac{-16 \pm \sqrt{16^2 - 4 * (-16) * (3)}}{2 * (-16)}$$

Multiply $4 * (-16) * 3$ to get -192, take $2 * -16$ to get -32, and compute $16 * 16$ to get 256. Plugging those calculations into the equation gives us the following:

$$t = \frac{-16 \pm \sqrt{256 - (-192)}}{-32}$$

Now, calculating $256 - (-192)$ and plugging in the result gives us:

$$t = \frac{-16 \pm \sqrt{448}}{-32}$$

©Brian Gillispie, 2016.

Using a calculator, $\sqrt{448} \approx 21.17$. Plugging[92] that in gives us:

$$t = \frac{-16 \pm 21.17}{-32}$$

Now, we have to evaluate the $\pm$ part of the formula. The plus/minus means to do the entire problem twice, once by doing a $+$ step here, and once by doing a $-$ step here. Because of this, we will break this problem up, as follows:

$$t = \frac{-16 + 21.17}{-32}, t = \frac{-16 - 21.27}{-32}$$

Notice this means we have two solutions for the quadratic equation, due to the evaluation of the $\pm$ step. Calculate both solutions completely to get the following for our two solutions:

$$t = 1.16, \quad t = -0.1646875$$

This means that the object hit the ground twice! The first time was about -0.16 seconds after the object was thrown, and the second time was about 1.16 seconds after the object was thrown. However, one of those answers makes no sense, because saying that the object the ground -0.16 seconds after it was thrown means it hit the ground before it was tossed into the air! Therefore, we have to discard the negative answer from our solutions, and conclude that the object hits the ground about 1.16 seconds after it was thrown into the air.

In the last Example the equation was already set up for us. In reality, this is rare, and the equation has to be created before the solution can be found, which we will show in the next Example.

---

[92] This will introduce rounding error into our solutions. Thankfully since from here all we will do is add then divide by 2a, the error will should not be too big.

©Brian Gillispie, 2016.

Example 7.1.4: An object is launched into the air with an initial velocity of 64 ft/sec, from an initial height of 32 feet. Assuming that $g = 32$ ft/sec^2, when does the object hit the ground?

Solution to 7.1.4: This time we have to create the equation before we can use the quadratic formula. In this problem our given value for g is 32, our initial velocity (v_0) is 64, and our initial height (h_0) is 32. Plugging those into our model gives us the following:

$$y = -16t^2 + 64t + 32$$

Now, to find when the object hits the ground we use the quadratic formula. In this problem, a is -16, b is 64, and c is 32. Plugging those numbers into the quadratic formula gives us the following to solve:

$$t = \frac{-64 \pm \sqrt{64^2 - 4*(-16)*(32)}}{2*(-16)}$$

Solving all of this using the order of operations gives us the following:

$$t = \frac{-64 \pm \sqrt{6144}}{-32}$$

Which, simplifies to the following:

$$t = 4.45, -0.45$$

However, an object cannot hit the ground in negative time, so the only answer is $x = 4.45$. Therefore, the object hits the ground after 4.45 seconds have passed.

We can also use the height equation and the quadratic formula to compute when and if an object ever reaches a given height. To do this, plug in the desired height in for y, and then solve the new equation using the quadratic formula. Let's demonstrate how to do this now.

©Brian Gillispie, 2016.

Example 7.1.5: Using the model from Example 7.1.4, which was $y = -16t^2 + 64t + 32$, how long until the object reaches a height of 75 feet? Or does the object reach a height of 75 feet ever?

Solution to 7.1.5: To solve this problem, we have to plug in 75 for y in our equation, then solve for t using the quadratic formula, once we set the equation equal to zero. Plugging in 75 for y gives us:

$$75 = -16t^2 + 64t + 32$$

However, to use the quadratic equation, this equation needs to be equal to zero first. To make the equation equal to zero, we will subtract 75 from both sides of the equation. Doing this yields the following:

$$0 = -16t^2 + 64t - 43$$

Now we can use the quadratic formula on this equation, with a equal to -16, b equal to 64, and c equal to -43. Plugging those values into the quadratic formula gives us:

$$t = \frac{-64 \pm \sqrt{64^2 - 4*(-16)*(-43)}}{2*(-16)}$$

Simplifying everything inside the square root (and some other places) gives us:

$$t = \frac{-64 \pm \sqrt{1344}}{-32}$$

Which, simplifies to the following:

$$t = 0.85, 3.15$$

This says the object will reach a height of 75 feet twice, once after 0.85 seconds pass, and once after 3.15 seconds pass. This makes sense as the object will hit the desired height once while going up into the air, and once while coming back down. Therefore, we would say that the object will reach a height of 75 feet at two different times, 0.85 seconds after the object is launched into the air, and 3.15 seconds after the object is launched into the air.

Example 7.1.6: Using the same model as we used in the last example, when does the object reach a height of 100 feet? Or, does the object ever reach that height?

Solution to 7.1.6: Using the same model as last time means we are using the equation $y = -16t^2 + 64t + 32$, and this time we want to know can our y value (height) ever be 100? To answer that, plug in 100 for y, and solve for t. Doing this gives us:

$$100 = -16t^2 + 64t + 32$$

Subtract both sides by 100 so we can use the quadratic formula. Doing this gives us:

$$y = -16t^2 + 64t - 68$$

Now we can use the quadratic formula, with a equal to -16, b equal to 64, and c equal to -68. Plugging these numbers into the quadratic formula gives us:

$$t = \frac{-64 \pm \sqrt{64^2 - 4 * (-16) * (-68)}}{2 * (-16)}$$

Evaluating everything inside the square root (and a couple other places) gives us:

$$t = \frac{-64 \pm \sqrt{-256}}{-32}$$

©Brian Gillispie, 2016.

However, we cannot compute a negative square root and end up with a real number answer. Therefore, since there are no real number solutions to the quadratic formula, this means that this problem has no solution. Therefore, we can conclude that this object will never reach a height of 100 feet, since no solutions exist to the corresponding quadratic equation.

While the method used in Example 7.1.6 works for finding if an object reaches a maximum height, it can be a tad tedious, especially if all we want to know is if the object can reach the height or not. Turns out, there is another method we can use to determine if an object reaches a certain height, and this is based on what we know of objects thrown into the air. All objects that are tossed into the air fly up until they reach a maximum height[93] then they fall back down. Therefore, if we could find this maximum height, we then could use it to determine if an object can reach a specific height quickly.

7.1.3: Maximums or Minimums of Quadratic Models

All quadratic models have an input which returns either a maximum or a minimum value for the model. This point of input can be found via the following method:

> *Definition 7.1.2:* For the quadratic $y = at^2 + bt + c$, the t value that yields the maximum or minimum value is the value:
>
> $$t = \frac{-b}{2a}$$
>
> If $a < 0$, then this point will yield a maximum. If $a > 0$, then this point will yield a minimum

It is important to remember that the value you find here only tells you what input yields the maximum or minimum. It does not tell you what the maximum or minimum value is. To find

---

[93] Or they reach escape velocity, but we'll assume you aren't throwing things hard enough for that to be possible.

©Brian Gillispie, 2016.

that, we will still need to plug the x value this formula finds into the equation. Let's demonstrate this now on our model Examples 7.1.4.

Example 7.1.7: What is the maximum height reached by the object modeled by the quadratic equation $y = -16t^2 + 64t + 32$?

Solution to 7.1.7: Based upon Definition 7.1.2, an input of $\frac{-b}{2a}$ will give us our maximum height. In this equation, a is equal to -16, and b is equal to 64. Plugging those into the formula tells us that a maximum will occur when:

$$t = \frac{-64}{2*(-16)} = 2$$

So if t is 2, then the object will be at its maximum height. Plugging in 2 for t into the original quadratic equation gives us that $y = 96$. Therefore, the maximum height this object can reach is 96 feet.

Example 7.1.8: A baseball pitcher hurls a baseball into the air as hard as he can, with a velocity of 66 ft/sec, from an initial height of 6 feet. If we assume that $g = 32$ ft/sec², what is the maximum height this baseball reaches?

Solution to 7.1.8: This time we need to set up our equation first. We are given that g is 32 ft/sec², the initial velocity (v_0) is 66, and the initial height (h_0) is 6. Plugging those in gives us the following for our model:

$$y = -16t^2 + 66t + 6$$

Now we want to find out the maximum height the baseball reaches. In this equation, a is equal to -16, and b is equal to 66. Plugging those into the formula for finding the maximum height gives us that the maximum height will occur when:

©Brian Gillispie, 2016.

$$t = \frac{-66}{2*(-16)} = 2.0625$$

In other words, the baseball will reach its maximum height after 2.0625 seconds pass. To find out what the maximum height is, we need to plug in 4.125 into the equation for t. If we do that, we get $y = 74.0625$ feet. Therefore, we conclude that the baseball reaches a maximum height of around 74.0625 feet.

©Brian Gillispie, 2016.

7.1 Exercises

For Exercises 1 - 8, create the height equation with the given information. Do not attempt to do anything else with the equation.

1. Initial velocity is 32 ft/sec, initial height is 80 feet, $g = 32$ ft/sec^2

2. Initial velocity is -16 ft/sec, initial height is 100 feet, $g = 32$ ft/sec^2

3. Initial velocity is 250 ft/sec, initial height is 2 feet, $g = 32$ ft/sec^2

4. Initial velocity is 600 ft/sec, initial height is 50 feet, $g = 32$ ft/sec^2

5. Initial velocity is 0 ft/sec, initial height is 1500 feet, $g = 32$ ft/sec^2

6. Initial velocity is 44.8 ft/sec, initial height is 4.2 feet, $g = 32$ ft/sec^2

7. Initial velocity is 25.6 ft/sec, initial height is 5.1 feet, $g = 2.66$ ft/sec^2

8. Initial velocity is 200 ft/sec, initial height is 0 feet, $g = 40$ ft/sec^2

9. A baseball is hurled into the air as hard as you can throw it. Given that the initial velocity of the baseball is 52 ft/sec, and the initial height is 6 feet, if $g = 32$ ft/sec^2, when does the baseball hit the ground?

10. A beanbag is dropped from the top of a building. If the initial velocity of the beanbag is 0 ft/sec, and the initial height is 120 feet, if $g = 32$ ft/sec^2, when does the beanbag hit the ground?

11. Repeat Exercise 10, but this time let $g = 2.66$ ft/sec^2

©Brian Gillispie, 2016.

12. You lob a football towards your friend. If the football has an initial velocity of 42 ft/sec, and an initial height of 4.9 feet, if $g = 32$ ft/sec^2, when does the football hit the ground?

13. For the situation in Exercise 12, you are trying to throw the football at least 10 feet in the air. When does the football reach a height of 10 feet? Or does it?

14. While on the moon, you decide to toss a rock into the air as hard as you can. If the rock is thrown with a force of 75.6 feet/sec, and an initial height of 4 feet, if $g = 5.3$ ft/sec^2, what is the maximum height reached by this rock?

15. For the situation in Exercise 14, if you decide to toss this same rock on another planet where $g = 80$ feet/sec^2 instead, what is the maximum height reached by the rock?

16. A paperback book is knocked off a ledge. If the book is knocked off the ledge with an initial velocity of -3 ft/sec, an initial height of 10 feet, if $g = 32$ ft/sec^2, when does the book hit the ground?

17. Does the book in Exercise 16 ever reach a height of 11 feet? Why or why not?

18. You decide to throw a rock into a crater as hard as you can. If you throw the rock with an initial velocity of -80 ft/sec, and at an initial height of 1200 feet, if $g = 6.72$ ft/sec^2, when does the rock hit the bottom?

19. For the data in Exercise 18, what is the maximum height ever reached by the rock?

20. Repeat Exercise 18, but this time assume $g = 45$ ft/sec^2

©Brian Gillispie, 2016.

7.2: Polynomial Interpolation

In the last Section, we discussed how to use a well-known quadratic model. However, what we did not do was discuss how to create this (or other) quadratic models from a few known data points. In this Section, we are going to show how we can create a quadratic (and other) models, using what is known as polynomial interpolation.

The idea behind polynomial interpolation is we have a few data points, and wish to find a polynomial that passes through all of our given data points. In order to find such a polynomial, we are going to have to solve what is known as a system of equations. Usually solving a system of equations is done with technology, and most graphing calculators and spreadsheet programs have built in functions that will allow you to solve such a system. In this Section of the book, we will assume the reader has access to such a program and is using it to solve the systems of equations that arise while creating a polynomial. If you need it, directions for the TI-83 and TI-84 brand calculator can be found in the end of the Section. If you use another calculator, we have to refer you to your user's manual[94].

To see how polynomial interpolation works, let's look at a problem that we can solve using techniques from earlier in this book, and solve it instead with our polynomial interpolation method. For our problem, we have been observing some mice in the field, and have discovered that on week 1 there are 15 mice, and on week 2 there are 20 mice. This gives us two points $(1, 15)$ and $(2, 20)$. As we saw previously, we could create a line to model these data points, by finding the slope and y intercept with this information. This time, we are going to create the line via the method of polynomial interpolation. We invite the reader to try to find the line that fits through these points now though, and compare their answer with ours once we are done.

To create our polynomial that fits the line, we need to start with a general equation of a line. Recall that the equation of a line is of the form $y = mx + b$, and we need to find out what m and b are for this line so that it will pass through our two given data points. To do that, we need to plug in our given points in for x and y. Doing that gives us the following two equations:

---

[94] This is because every calculator has its' own method of setting up systems of equations. Even two calculators in different models but same company will often have different directions, making it impossible for us to include every set of directions in the book.

©Brian Gillispie, 2016.

$$15 = 1m + b$$
$$20 = 2m + b$$

This is known as a system of two equations, with two variables, the two variables being m and b. To solve this, we solve the equation $15 = m + b$ for one of the variables. Which variable does not matter, so we will solve this equation for b. Doing that gives us the new system of equations:

$$15 - m = b$$
$$20 = 2m + b$$

Now, plug our solution for b from the first equation into the second equation. If we do that, we get the following:

$$15 - m = b$$
$$20 = 2m + (15 - m)$$

Notice how $15 - m$ took the place of b in the second equation. This step is called substitution, and can be legally done with anything that is equal to the original variable. Now, the second equation can be solved for m, which, when we do that, we get that $m = 5$. So now, our system has become:

$$15 - m = b$$
$$m = 5$$

Since we know the solution to m now, we plug it back into the first equation, which gives us that b was 10. This makes our solution for the variables the following:

$$b = 10$$
$$m = 5$$

©Brian Gillispie, 2016.

And, with that, our final equation for the line is:

$$y = 5m + 10$$

Now, it is true that this was not easier than finding the equation of the line via the earlier methods. In practice, if we have two points, modelers often use the methods of the earlier Chapters instead, as they are much quicker than the method we used here. However, if we have three or more points, then we have to use this or a similar polynomial interpolation method[95] to find the desired polynomial.

Before we continue though, we do need to bring up an important point. As the reader can see, using polynomial interpolation via this method can be long. It is **highly** recommended that the reader have a calculator or mathematical program that can solve systems of equations while working this section. Otherwise, these problems will be very time consuming, and near impossible to solve accurately. From my observations while teaching, when solving systems of equations by hand, it takes on average five minutes to solve a system of two equations, two variables, twenty minutes to solve a system of three equations, three variables, almost an hour to solve a system of four equations, four variables[96], and so on. The time increases quickly as the number of equations increases. For this reason, I will be using a TI-84 to solve all remaining problems in this Section, and I encourage the reader to use something similar.

With that said, we will now summarize the technique for polynomial interpolation, for making a polynomial of degree n with $n + 1$ data points.

Polynomial Interpolation:

To create a polynomial $y = a_n x^n + \cdots + a_1 x + a_0$, with $n + 1$ data points known, we have to take those data points, plug them in for x and y, and then solve the resulting system of equations for the coefficients $a_0, a_1, \ldots, a_n$.

---

[95] There are other interpolation methods. See [13] for more on this.
[96] When I was new at teaching, a class convinced me to work a system of four equations, four unknowns in class as an example. It took the entire 50 minute class period.

©Brian Gillispie, 2016.

Let's demonstrate this method on a few Examples now. Remember, for these Examples, the systems shown will have been solved via a calculator or mathematical software!

Example 7.2.1: A ball is tossed into the air. After zero seconds have passed, the ball is zero feet off the ground. After two seconds have passed, the ball is 64 feet off the ground, and after four seconds have passed, the ball is zero feet off the ground. Create a polynomial model for this situation.

Solution to 7.2.1: First of all, we need to figure out what our points are. In this model, if we let x represent seconds and y represent height, then our data points are $(0,0)$, $(2,64)$ and $(4,0)$. We have three points, which means we can create a quadratic model by Definition 7.2.1. Therefore, our model will be of the form $y = ax^2 + bx + c$. Plug in our three points into this equation for x and y, and we end up with the following three equations:

$$0 = c$$
$$64 = 4a + 2b + c$$
$$0 = 16a + 4b + c$$

Solving this with software gives us the following solutions to a, b, and c:

$$a = -16$$
$$b = 64$$
$$c = 0$$

Which, gives us a model of:

$$y = -16x^2 + 64x$$

Example 7.2.2: Sales over a four week period are as follows: On week one there were $3000 in sales. On week two there were $3600 in sales. On week three, there were $4300 in sales. On week four, there were $5000 in sales. Create a polynomial model.

Solution to 7.2.2: Again, we need to first figure out what our points are, and how many of them we have. Start off by letting x represent the week and y representing the sales. This then gives us the points $(1, 3000)$, $(2, 3600)$, $(3, 4300)$ and $(4, 5000)$. We have four data points, which means we can create a third degree polynomial. Therefore, our model will be of the form $y = ax^3 + bx^2 + cx + d$. Plug in our points then for x and y, which will give us the following system of four equations:

$$3000 = a + b + c + d$$
$$3600 = 8a + 4b + 2c + d$$
$$4300 = 27a + 9b + 3c + d$$
$$5000 = 64a + 16b + 4c + d$$

Using software to solve this system gives us that the solution for a, b, c, and d is:

$$a = 16.\bar{6}$$
$$b = 150$$
$$c = 266.\bar{6}$$
$$d = 2600$$

This gives us our final model of:

$$y = -16.\bar{6}x^3 + 150x^2 + 266.\bar{6}x + 2600$$

Just like with our previous models, we can use these models to make some predictions as well. However, sometimes in the process of making the prediction we will run into an equation

©Brian Gillispie, 2016.

we cannot mathematically solve. When this happens we have to resort to the method of guess and check instead[97]. Let's demonstrate this by trying to calculate a couple of prediction now.

Example 7.2.3: Using the model found in Example 7.2.2, predict how many sales there will be in week 6?

Solution to 7.2.3: Our model for Example 7.2.2 was $y = -16.\overline{6}x^3 + 150x^2 + 266.\overline{6}x + 2600$ with x representing weeks, and y representing sales. Therefore, to find how many sales will be made in week 6, we need to plug in 6 for x, and calculate. If we do that, our equation becomes:

$$y = -16.\overline{6} * 6^3 + 150 * 6^2 + 266.\overline{6} * 6 + 2600 = 6000$$

Therefore, this model predicts that on week 6 we will have 6,000 in sales.

Example 7.2.4: Using the model in Example 7.2.2, predict when the business will have $7000 in sales.

Solution to 7.2.4: Our model for Example 7.2.2 has x defined in weeks, and y defined in sales. This means we have to plug in $7000 for y then, and therefore have to solve the following equation:

$$7000 = -16.\overline{6}x^3 + 150x^2 + 266.\overline{6}x + 2600$$

Unfortunately, there is no consistent method for solving cubic equations. Therefore, the best way to solve this problem is to either guess and check, use a zero finding method[98], or use excel (or other software) to create a data set and see when and if the data ever crosses 7000. Using software to generate a table gives us the following table:

---

[97] Or you can use a method covered in Section 9.1 instead.
[98] A couple zero finding methods are covered in Chapter 9.

©Brian Gillispie, 2016.

Week	Sales
1	3000
2	3600
3	4300
4	5000
5	5600
6	6000
7	6100
8	5800
9	5000
10	3600
11	1500
12	−1400
13	−5200

Based on this table, it appears that sales never reach 7000, therefore we conclude that sales will never get that high, for any week.[99]

This last example illustrates the biggest disadvantage of higher degree polynomial models, which is if we wish to find out which x value returns a specific y value, it is near impossible to do it short of guessing and checking. For this reason, many modelers prefer to use second degree polynomials or lower, just because those are easier to solve for x when the situation requires it.

In addition, high degree polynomial models have the issue that the values between sequential data points can vary wildly. It is very possible to create a polynomial model that predicts that in week 6 a store will sell 99,000 copies of a book, in week 7 the store will sell −15,000 copies of the same book, and in week 8 the store will sell 278,901,781 copies of the book. This occurs because in polynomials of high degree, the values in the polynomial tend to fluctuate wildly. However, since very few things in the real world fluctuate that wildly, modelers tend to stay away from very high degree polynomials in modeling for that reason.

As a result of the issues mentioned here, polynomial models of higher than third degree are rarely created in practice, and models of third degree are often used only for situations when

---

[99] Interestingly enough, sales also went negative by week 12. Normally we would report this as 0 sales, but left the raw computed numbers for this Example.

©Brian Gillispie, 2016.

we know x and wish to predict y, due to the difficulty in solving for x in third degree polynomials. What this means then is if a modeler has more than four data points, they often either use a spline model, find a low degree polynomial that comes closest to the points and keep the error as low as possible, or they pick four (or less) points for the purpose of creating the model and hope the error is low for the rest of the data. All of these methods have advantages and disadvantages, and it is up to the modeler to decide which method they think works best for their situation.

Appendix to Section 7.2: Calculator Directions

In our work in Section 7.2, we had to use software to solve our systems of equations. While it is not possible to give comprehensive directions to every software program or calculator out there, we will list the directions here for two of the more popular calculators as of the time this Section was written. The directions here are for a TI-84 calculator, and should need very little adjustment to work for a TI-83 brand calculator as well.

To solve a system of equations with a calculator, you have to first convert the system to a matrix. When converting to a matrix, you only post the coefficient for each variable, not the variable itself. For example, if our system was the following:

$$c = 0$$
$$4a + 2b + c = 64$$
$$16a + 4b + c = 0$$

Our matrix would be the following for this system. Notice that coefficients for variables that are not present are listed as zero, and that the first column corresponds to coefficients of a, the second column is the coefficients of b, the third column the coefficients of c, and the fourth column the numbers on the other side of the equal sign[100].

$$\begin{bmatrix} 0 & 0 & 1 & 0 \\ 4 & 2 & 1 & 64 \\ 16 & 4 & 1 & 0 \end{bmatrix}$$

To enter this matrix into the calculator, hit the 2nd button, then select Matrix. Scroll over to the Edit button, then select a letter (remember which letter you selected!), and hit enter. It will then ask you for the size of the matrix. Size is always reported as rows by columns, so in this example we have three rows, and four columns, and would enter 3 by 4. Then enter each number in the matrix individually. Double check that all numbers are entered correctly, and once you are

---

[100] While you could do c, then b, then a in the matrix, you *must* always have the numbers on the other side of the equal sign last in the matrix. Doing it any other way will confuse the software and the answers will be incorrect.

©Brian Gillispie, 2016.

satisfied that all entries are correct, hit 2nd, then Quit. This will exit you out of the matrix entering field.

Now, to have the calculator solve the matrix, hit the 2nd button, then select Matrix again. This time though, scroll over to the Math button. Scroll down until you find the commend rref, on a TI-84 it is option B so some scrolling down will be needed. Hit enter.

Now, hit the 2nd button again, then select Matrix. Stay on the Names tab this time, and select the matrix you entered all the data into. Hit enter once you are on that matrix. Then, put a closing parenthesis on your entry, and hit enter, and the calculator should return another matrix. If you do these steps on the Example matrix listed above, the calculator will return:

$$\begin{bmatrix} 1 & 0 & 0 & -16 \\ 0 & 1 & 0 & 64 \\ 0 & 0 & 1 & 0 \end{bmatrix}$$

Now, to interpret this answer, you need to remember which column was for which variable. In our example, our first column was for variable a, the second column was for b, and the third was for c. The last column is always the number after the equal sign. This translates to tell us that our system of equations has the following answer:

$$a = -16$$
$$b = 64$$
$$c = 0$$

Which, was what we got when we worked this system of equations back in Section 7.2

©Brian Gillispie, 2016.

7.2 Exercises

For all of these Exercises, create the appropriate polynomial model.

1. A car is initially observed doing 55 mph, then 4 seconds later the car is doing 70 mph.

2. Sales of the new Chocolate Fish sandwich were 2 on May 2nd, and 4 on May 3rd.

3. A ball is tossed into the air. When the ball is released, it is zero feet in the air. After 1 second has passed, the ball is 72 feet in the air. After two seconds have passed, the ball is 112 feet in the air.

4. A punter punts away the football to the other team. When the ball is kicked it is 1 foot off the ground. One second later, the football is 19.14593 feet off the ground, and two seconds later, the football is 5.291859 feet off the ground.

5. Using your model in Exercise 4, predict when the football will hit the ground.

6. Using your model in Exercise 4, calculate the maximum height reached by the football.

7. While observing ducks on a pond, you record the following counts. On August 2nd there were 8 ducks on the pond, on August 3rd there were 6 ducks on the pond, on August 4th there were 10 ducks on the pond, and on August 5th there were 5 ducks on the pond.

8. Using your model for Exercise 7, predict how many ducks there will be on week 5. If you get a negative answer, report 0 ducks.

9. Ticket sales for a football team are as follows: On opening day (aka game 1) 72,657 tickets are sold. For the second game, 65,892 tickets are sold. For the third game, 66,782 tickets are sold, and for the fourth game, 72,657 tickets are sold.

©Brian Gillispie, 2016.

10. For the model in Exercise 9, if 72,657 tickets sold means the game was sold out, use your model to see if any of the remaining 3 home games are sold out. *Hint: Treat answers of more than 72,657 as 72,657.*

11. You are assigned to Antarctica, and are in charge of observing penguins there. When you arrive, you count 156 penguins on December 10th. On December 11th, you count 200 penguins. On December 12th, you count 192 penguins. On December 13th, you count 212 penguins, and on December 14th you count 250 penguins.

12. Use your model in Exercise 11 to predict how many penguins will be present on December 15th. If you get a negative answer, report it as zero penguins.

13. Use your model in Exercise 11 to predict how many penguins will be present on December 18th. If you get a negative answer, report it as zero penguins.

14*. Use your model in Exercise 11 to predict when there will be 500 or more penguins present.

©Brian Gillispie, 2016.

7.3: Proportionality

In this Section, we will discuss another type of continuous model, and that is the proportionality model. Proportionality models exist when two or more variables are related in a certain way such that when we increase one variable, the other variable or variables have to change as well.

7.3.1: Direct Proportionality

First, let's consider the case where increasing one variable in the equation forces us to increase the other as well. This type of proportionality is called direct proportionality[101], which can be modeled as follows:

> *Definition 7.3.1:* Variable y is **directly proportional** to x if whenever values of x increase, y also increases. This is represented as the following, with k the constant of proportionality:
>
> $$y = kx$$

In practice, the value of k is not known, and has to be found by using information provided. Once k is known, we then use that information to find other, unknown information. Let's demonstrate with an Example.

Example 7.3.1: If y is directly proportional to x, and when y is 10, x is 20, find y when x is 40.

Solution to 7.3.1: Before doing anything, set up the equation. In this problem, we know y is directly proportional to x. This tells us that the equation is:

$$y = kx$$

---

[101] Some books use the term variation instead of proportional.

©Brian Gillispie, 2016.

We also know that when y is 10, x is 20. That means that we need to plug both of these givens into the equations at the same time. If we do that, we can then find k, which we need to answer the asked question. Plugging in $y = 10$ and $x = 20$ into the equation at the same time gives us:

$$10 = k(20)$$

Solve for k by dividing both sides by 20 to find that:

$$k = 0.5$$

With k found, we update our model with the known value of k. This gives us an updated model of:

$$y = 0.5 * x$$

Now we can answer the desired question. We want to know what y is, when x is 40. Plug in 40 for x into the equation to find that y is:

$$y = 0.5 * 40 = 20$$

Therefore, when x is 40, y is 20.

It is also possible to have two variables set up in such a way that one variable is directly proportional to the square or the cube or even the square root of the other variable. When that happens, we set up the equation as normal, being sure to apply the math operation to the right variable. Let's demonstrate that now with an Example.

©Brian Gillispie, 2016.

Example 7.3.2: It is often believed that the length of the skid marks a car leaves behind (in feet) is directly proportional to the square of the speed of the car (in mph)[102]. If a car going 30 mph leaves behind 40 feet skid marks, how long would the skid marks of a car going 60 mph be?

Solution to 7.3.2: This example is a little different than our previous Example, as the variable length of skid marks is directly proportional to the square of the speed of the car. Nevertheless, we can still use our rules for directly proportional problems to solve this problem. To do this, look at what the given statement is saying first. The statement is saying that whenever we increase the speed, then square the speed, we still increase the length of the skid marks. Therefore, using the definition of directly proportional, with x^2 instead of x, gives us the following model:

$$y = kx^2$$

Note here that y stands for length of the skid marks, and x stands for the speed of the car. Now, we need to find k. To find k, we know that when the car is going 30 mph, the skid marks are 40 feet long. Plug in the given values into the model to get:

$$40 = k * 30^2$$

Solve this for k to get:

$$k = \frac{4}{90}$$

With k found, update the model, so that we can use it to answer the desired question. Updating the model gives us:

$$y = \frac{4}{90}x^2$$

---

[102] This model is often used in accident reconstruction to try to figure out how fast the car was really going, and if that speed matches the reported speed.

©Brian Gillispie, 2016.

Now, we want to know how long the skid marks should be if the car was going 60 mph. Plugging in 60 for x gives us:

$$y = \frac{4}{90} * 60^2 = 160$$

This means that if the car was going 60 mph, the skid marks should be about 160 feet long.

7.3.2: Inverse Proportionality

Next, let's discuss the case where when we increase one variable, the other variable decreases. This type of proportionality is called inverse proportionality, and can be modeled by the following:

> *Definition 7.3.2:* Variable y is **inversely proportional** to x if whenever we increase x, we decrease y. This is modeled as follows:
>
> $$y = \frac{k}{x}$$
>
> Where k is the constant of proportionality

Inverse proportionality problems are solved the exact same way as direct proportionality problems, as we will now demonstrate:

Example 7.3.3: The time it takes to arrive at a destination varies inversely with the speed that one is traveling at. If it takes you 4 hours to drive to a location when you are going 45 mph, how long will it take you to drive to this location if you are going 35 mph?

Solution to 7.3.3: Start this problem by creating the equation. We know that time varies inversely with the speed we drive at, so we will let y be time and x be speed. This means that our equation for this situation is:

$$y = \frac{k}{x}$$

Next, we use the given information to find k. We know that when we drive 45 mph it takes 4 hours, so if we plug those numbers into the model, we end up with:

$$4 = \frac{k}{45}$$

Finish solving the equation, and you will learn that k is 180. Update the model with this known value of k, which gives us that the model is:

$$y = \frac{180}{x}$$

We can now answer the question asked. We want to know how long it will take to reach our destination if we drive 35 mph. Plugging in 35 mph into the equation, we end up with:

$$y = \frac{180}{35}$$

Which simplifies to 5.14 hours to reach our destination, at 35 mph.

One of the more common applications of inversely proportional variables is in modeling light and sound. Light and sound follow what is known as an inverse square law, which state that the intensity varies inversely as the square of the distance, which is modeled by the following with I for intensity and d for distance:

$$I = \frac{k}{d^2}$$

Let's now work a few Examples using this inverse square law to model light and sound.

Example 7.3.4: The intensity of a light source (in watts) varies inversely as the square of the distance you are from the light source. The sun's light lands on Mercury with an intensity of 9126 watts per square meter, and Mercury is 0.387 AU from the Sun. How intense is the light on Earth, if the Earth is 1 AU from the Sun?

Solution to 7.3.4: Start out by creating the equation. In this problem, we know that light varies inversely as the square of the distance. This gives us the following equation, with I representing the intensity in watts, and d the distance in AU:

$$I = \frac{k}{d^2}$$

We also know that the intensity of the light is 9126 watts when the distance is 0.387 AU. Plug those in to find k, as follows:

$$9126 = \frac{k}{0.387^2}$$

Which, if we compute 0.387^2, then multiply both sides by that result, we find that:

$$1366.79 \approx k$$

©Brian Gillispie, 2016.

With k found, we can now update our model, which gives us:

$$I = \frac{1366.79}{d^2}$$

Now, we want to know how intense the light is on Earth, which is 1 AU away from the sun. Plugging in 1 for d gives us:

$$I = \frac{1366.79}{1^2}$$

Which calculates to give us that $I = 1366.79$. This means that the intensity of the light on the Earth from the Sun is about 1366.79 watts per square meters.

This last problem demonstrates something interesting. Notice how Earth was 1 AU away, and the intensity on Earth was exactly equal to the constant of proportionality. This is not a coincidence. It turns out that when we are modeling intensity, the constant of proportionality will always be how intense the item we are measuring is one unit of distance away from the object creating the intensity. In other words, if we measure our model in meters, then the constant of proportionality will always tell us how intense the light is one meter away from the light source. If we were modeling sound in miles, then the constant of proportionality would reveal how intense the sound is one mile away from the sound.

Let's demonstrate this on another Example:

Example 7.3.5: The intensity of a sound wave in watts is inversely proportional to the square of the distance someone is standing from the sound wave. A really loud explosion has occurred in a blasting zone. If the sound wave emitted a force of 0.5 watts to a worker who was standing 40 meters away, how loud was the sound to a worker who was only 1 meters away from the blast?

Solution to 7.3.5: Start out by creating the equation. In this case, we know that the intensity of the sound wave is inversely proportional to the square of the distance one is from the sound wave, which means that the model is:

©Brian Gillispie, 2016.

$$I = \frac{k}{d^2}$$

Next, we need to find k, which will also tell us how loud the sound is to someone only one meter away from the blast, as our distance is measured in meters. We know that the sound wave had an intensity of 0.5 watts to someone who was 40 meters away. Plugging those numbers into the equation gives us the following:

$$0.5 = \frac{k}{40^2}$$

This simplifies to $k = 800$, which means that someone who was only one meter away from the blast had the sound wave hit them with an intensity of 800 watts[103].

Example 7.3.6: You decide to deck out your home entertainment system with some really powerful speakers, then decide to crank up the speakers as loud as they can go (we highly recommend you do not really ever do this!). Assuming that at a distance of 1 meter, the sound wave has an intensity of 1000 watts, how loud is the sound wave to your next door neighbor, who is seated in his house, 30 meters away from your speaker? Use the rule that sound intensity varies inversely as the square of the distance from the sound to solve this equation.

Solution to 7.3.6: Start off by setting up the equation. Again, the sound intensity varies inversely as the square of the distance, which gives us:

$$I = \frac{k}{d^2}$$

Next, plug in what we know. We know that at a distance of 1 meter, the speakers are creating a sound wave of 1000 watts. Since k is the intensity of the sound at a distance of one unit away, we can plug in our given wattage for k this time. Doing that gives us:

---

[103] This is about 159 decibels, which means he also probably had hearing damage from the blast.

©Brian Gillispie, 2016.

$$I = \frac{1000}{d^2}$$

Next, we want to find out how intense this sound wave is to your neighbor, who is 30 meters away. Plugging in 30 for the distance gives us:

$$I = \frac{1000}{30^2}$$

Finishing the calculation tells us that the sound wave arrives to your neighbor with an intensity of 1.11 watts, approximately. This may not seem like a lot, but it is about 130 decibels still, which means that the sound wave from your movie hits him with about the same intensity as if he was at a rock concert. Better prepare for an angry knock at the door soon.

Being able to model with proportionality has a major advantage, and that is that the model can be created with only one data point. This allows the modeler to not have to spend extensive time collecting data before they have a model available to use. However, creating models with only one data point is dangerous, as you have no idea if the current trend continues or not, as you are just assuming it will. Therefore, it is highly recommended that proportionality is only be used for preexisting models that we already know follow the rules of proportionality, at least initially.

7.3 Exercises

For Exercises 1 - 10, use the given information to find k, then state what the final model will look like with k known.

1. y is directly proportional to x, and $y = 50$ when $x = 2$.

2. y is directly proportional to x, and $y = 100$ when $x = 900$

3. y is directly proportional to x, and $y = 2$ when $x = 1$

4. y is directly proportional to the square of x, and y is 10 when x is 75.

5. y is directly proportional to the cube of x, and y is 16 when x is 2.

6. y is inversely proportional to x, and $y = 10$ when $x = 2$

7. y is inversely proportional to x, and $y = 1000$ when $x = 10$

8. y is inversely proportional to the square of x, and $y = 50$ when $x = 4$

9. y is inversely proportional to the square root of x, and y is 100 when x is 25

10. y is inversely proportional to the square of x, and y is 6 when x is 50

Create the appropriate model, and solve the problem.

11. Skid marks are often believed to be directly proportional to the square of the speed of the car at the time they hit the brakes. If a car going 60 mph leaves 160 foot skid marks, how long would the skid marks be for a car going 80 mph?

12. Wages earned is directly proportional to the hours one works. If one makes $400 a week when they work 30 hours a week, how much would they make when they work 40 hours a week?

13. Taxes withheld are directly proportional to the wages one earns. If one had $100 in taxes withheld from a $1200 paycheck, how much in taxes would be withheld from a $2400 paycheck?

14. The mass of an object is proportional to its length cubed. If an object has a mass of 100 g when it is 5 cm long, how much mass would it have if it were 7 cm long?

15. The mass of a gorilla is proportional to their length cubed. If a 5.6 foot tall gorilla has a mass of 175 kilograms, how much mass would a 25 foot tall gorilla have?

16. Repeat Exercise 15, but this time figure out how much mass a 100 foot tall gorilla would have?

17. The intensity of sound is inversely proportional to the square of the distance from the sound wave. A speaker at a rock concert emits a sound wave that has intensity 1 watt when you are standing 50 feet from the speaker. How intense would the sound wave be if you were to stand 200 feet from the speaker?

©Brian Gillispie, 2016.

18. The intensity of light is inversely proportional to the square of the distance from the light source. A light emits 60 watts of illumination to a reader who is 6 feet from the blub. How much illumination would the light emit to someone who is standing 25 feet from the blub?

19. Repeat Exercise 18, but this time find out how much illumination the light would emit to someone standing 100 feet from the blub.

20. The intensity of sound is inversely proportional to the square of the distance from the sound. If a rock concert emits a sound wave that has intensity 10 watts to someone who is 50 feet from the closest speaker, how intense will the sound wave be to someone who is standing 500 feet away in the parking lot?

21. The zombie invasion has occurred, and you are trying to not make any sound as you sneak by an old base filled with zombies. However, you trip and say something. If the words you said had an intensity of $4.25 * 10^{-10}$ to someone who was 2 feet away from you, how loudly did your words sound to that zombie who was 15 feet away from you, assuming that the intensity of sound is inversely proportional to the square of the distance from the sound?

22. A dropped object makes a sound of $6.8 * 10^{-6}$ watts to someone who was 2 feet away from the object. How intense is the sound wave to someone who was standing 10 feet away from the dropped object, assuming that the intensity of sound is inversely proportional to the square of the distance from the sound?

©Brian Gillispie, 2016.

7.4: Trigonometric Modeling

The next continuous model we are going to explore and discuss is the trigonometric model. Trigonometric models are primarily used to model behavior that oscillates between a maximum and minimum value over a fixed period of time.

Before we begin our discussion of trigonometric models, we will first discuss some common terms used while working with trigonometric models.

7.4.1: Common Terms

In trigonometric modeling, we often wish to be able to represent our real world data by the following equation[104]:

$$y = A * \sin(B(x - c)) + D$$

However, before we can create such a model, we need to discuss what the various terms mean in the equation, so we can understand how they affect the model we will use. The four terms that we will need to define are phase shift, period, amplitude, and vertical translation. We will start with the definition of phase shift, which is below:

> *Definition 7.4.1:* The **phase shift** of the function $f(x) = \sin(B(x - C))$ is defined as the amount of horizontal translation that is applied to the graph of $g(x) = \sin(Bx)$, and the phase shift is equal to C.

This means that the phase shift is defined as by how much did the graph of the original function move to the right or left from the original. Notice also that the phase shift is the opposite

---

[104] The techniques in this section also work for cosine instead of sine, if you shift the model over appropriately. However, in this section, we will focus on using the sine model instead.

©Brian Gillispie, 2016.

of the sign of the original value in the model! Let's demonstrate this by finding a few phase shifts.

Example 7.4.1: Given the function $f(x) = \sin(2(x - 30))$, what is the phase shift of the function?

Solution to 7.4.1: In this function, the phase shift is the value being subtracted from x, which means the phase shift is 30.

Notice that the phase shift in the last Example is 30, and not -30. This is because the phase shift is always the number being subtracted from x. If the number is being added instead, rewrite it as a subtraction problem, as this next Example will demonstrate.

Example 7.4.2: Given the function $f(x) = \sin(x + 20)$, what is the phase shift of the function?

Solution to 7.4.2: In this function, the phase shift is the value being subtracted from x. However, we are adding 20, not subtracting it. This can be fixed by re-writing the equation as follows:

$$f(x) = \sin(x - (-20))$$

Now, we are subtracting negative 20, which means the phase shift of the function is -20

It is vitally important to remember that the phase shift is the value that is being subtracted from x. If there is a number being added instead, it is mandatory to re-write it as a subtraction problem of a negative number to find the phase shift.

Next, we will define the period of a trigonometric function. The period of any function is defined as the smallest value of p that makes this statement true, for all values of x:

$$f(x) = f(x + p)$$

©Brian Gillispie, 2016.

Thankfully, we will not need to solve that equation, as it has been proven that the period of $\sin(x)$ is 2π. This then means that we can define the period of $\sin(B(x - C))$ as follows:

> *Definition 7.4.2:* The **period** of the function $f(x) = \sin(B(x - C))$ is found as follows:
>
> $$period = \frac{2\pi}{B}$$

Notice that in order to find the period and phase shift properly, you must have x by itself. That means no $2x$, or $3x$, just x. If the equation starts with $2x$, you must factor the 2 out of all terms before proceeding. Let's demonstrate:

Example 7.4.3: Given the function $f(x) = \sin(2x - 30)$, what is the period and phase shift of this function?

Solution to 7.4.3: A common mistake on this question is to say the period is π and the phase shift is 30, however that only has one out of the two correct. To see why one of them is wrong, we need to remember that our equation has to be in the form $f(x) = \sin(B(x - C))$, but we don't have it in that form, as the x is not alone. To make the equation in that form, we must factor the expression $2x - 30$ by factoring an 2 out. If we do that, our function now becomes:

$$f(x) = \sin(2(x - 15))$$

Notice that when we factor the 2 out, the number we are subtracting x by also reduces. With the equation in this factored form, we can see that our phase shift is 15 for this function, not 30 like we first thought.

To find the period, we need to take 2π and divide it by the number we just factored out. If we do this, we find that our period is $2\pi/2$, or π.

Therefore, for this function, our phase shift is 15, and our period is π.

Next, we will define the amplitude of a function, which is defined as follows:

> *Definition 7.4.3:* The **amplitude** of the function $f(x) = A * \sin(B(x - c))$ is defined as how much the function oscillates from a fixed point, and is given by $|A|$

The amplitude, by definition, can never be negative. Let's now find the amplitude for a couple of functions.

Example 7.4.4: For the function $f(x) = 2 * \sin(x + 25)$, what is the amplitude for the function?

Solution to 7.4.4: The amplitude is defined as how much the function is oscillating from a fixed point (usually zero). In this case, the amplitude is given by $|2|$, which is 2. Therefore, the amplitude of our given function is 2.

Example 7.4.5: For the function $f(x) = -3 * \sin(4x + 40)$, find the amplitude, phase shift, and period.

Solution to 7.4.5: This equation is not in the form we need it in, so first thing is to make the equation in the desired form. Start by factoring the 4 out of the $4x + 40$ term, which gives us:

$$f(x) = -3 * \sin(4(x + 10))$$

Notice the 4 does not go outside of the function[105]. Also, we need to have x minus something, so we re-write the plus 10 term as follows:

$$f(x) = -3 * \sin(4(x - (-10)))$$

---

[105] In general, everything inside the function must stay inside the function until it is evaluated.

©Brian Gillispie, 2016.

Now the equation is in the desired form. The phase shift is the term we are subtracting from x, which is -10. The period is $\frac{2\pi}{4}$, which simplifies to $\frac{\pi}{2}$. And finally, the amplitude[106] is $|-3|$, which simplifies to give us 3.

The last term we need to define is the vertical shift, or vertical translation, depending on what you are used to calling it. This is defined as follows:

> *Definition 7.4.4:* The **vertical shift** of the function $f(x) = A * \sin(B(x - C)) + D$ is defined as how much we move up or down the fixed point the graph oscillates around, and is given by D

Let's find a few vertical shifts now before proceeding.

Example 7.4.6: Find the vertical shift of the function $f(x) = \sin(x) + 5$

Solution to 7.4.6: The vertical shift is how much we are adding (or subtracting) to the function, which in this case is 5.

Example 7.4.7: Find the vertical shift of the function $f(x) = -2 * \sin(3(x - 10)) + 10$

Solution to 7.4.7: Careful here, as a common wrong answer is zero. However, the minus ten is instead the function sine, and the plus ten is outside the function, so we cannot do that.

Instead, the correct answer is 10, as that is the number we are adding to the function. Therefore, we conclude that the vertical shift is 10 for our function.

---

[106] Be careful! Many people will say the amplitude is 12, from $-3 * 4 = -12$, then taking the absolute value of -12. This mistake is actually very common, and results from a misunderstanding of the distributive rule. The distributive rule states you can pull multiplication outside (), but only (), not functions! The distributive rule does not apply to functions, except in very rare cases.

©Brian Gillispie, 2016.

7.4.2: Creating Trigonometric Models

With the trigonometric terms now defined, we are ready to create a trigonometric model. To start this discussion, let's look at an example. Pretend that you are in charge of weather forecasting and wish to create a model that can take as input the month, and return as output the projected average high temperature of the region. You collected data over 3 month intervals, and your data set is as follows:

Month and Year	Average High Temp
December 2005	36
March 2006	62
June 2006	91
September 2006	64
December 2006	31
March 2007	59
June 2007	97
September 2007	75
December 2007	41

Notice that the data set seems to be oscillating between the temperatures in December and the temperatures in June. This makes the data set a good candidate for a trigonometric model.

In order to make our trigonometric model, we need to first figure out how often the data set oscillates. In this example, our data set oscillates every 12 months, so our period should probably be 12. In order to force our period to be 12, we need in our equation $A * \sin(B(x - C)) + D$, the value of B to satisfy:

$$\frac{2\pi}{B} = 12$$

This solves to give us:

$$B = \frac{\pi}{6}$$

©Brian Gillispie, 2016.

Next, we need to figure out by how much this function should oscillate. A reasonable assumption is any function that models this data set should oscillate between the highest data point in the set, and the lowest data point in the set. In this example, our highest data point is 97, and our lowest data point is 31. This means then that our amplitude should oscillate between 97 and 31, which, in order to make that happen, we need the following to hold:

$$\frac{97-31}{2} = A$$

Which, solves to give us for our amplitude:

$$A = 33$$

However, we are not done. If we just use the function:

$$f(x) = 33 * \sin(\frac{\pi}{6}(x))$$

We will be disappointed with the results of this function, as it will say our high temperatures oscillate between -33 and 33 degrees throughout the year. The reason this happens is the amplitude oscillates around a fixed point, and that fixed point defaults to zero unless we change it. To change this fixed point, we need to add a vertical shift to the function. We want our function to oscillate between 97 and 31, which means that we need the fixed point to be the average of these two numbers. This means that our vertical shift D should be:

$$D = \frac{97+31}{2}$$

Which, solves to give us for our vertical shift:

$$D = 64$$

©Brian Gillispie, 2016.

If we add in that term, our function will now oscillate between 31 and 97, like we want. However, we still have a problem, as the function will return the temperatures for the wrong months. This is because the function sin(x) starts out at the vertical shift, then goes up, then back down. If we use the function as stated so far, assuming that an input of zero means December, we will predict that the temperature will be 64 degrees in December, 97 degrees in March, 64 degrees in June, and so on. The values are all off by 3 months. To fix this, we need to add in a phase shift. However, we need to be careful how we add this in. When we add in our phase shift, we need to move the graph over three months to the right, which translates to a phase shift of 3. This means that our final function needs to be:

$$f(x) = 33 * \sin\left(\frac{\pi}{6}(x-3)\right) + 64$$

The reader should verify[107] that this model does indeed now oscillate between 31 degrees in December and 97 degrees in June, as desired.

This process can be applied in general, provided the data set has equally spaced oscillations. If the data set does not have equally spaced oscillations, like in the data set 61 98 61 75 98 75 61 72 78 98 78 72 98, then a trigonometric function will not work well. It is important to verify that the oscillations are equally spaced, or nearly equally spaced, before using a trigonometric function. Also, the way we defined our function, it is important to note that $sin(x)$ is defined in radian mode. Attempting to use degree mode to calculate the answers will result in inaccurate answers, though those could be fixed by using 360 instead of 2π when finding the period.

With that said, the general form for a model for equally spaced oscillatory data set can be found as follows:

---

[107] Make sure you are in radian mode though! If you treat x as meaning degrees, the answers will still be wrong.

©Brian Gillispie, 2016.

> *Definition 7.4.5:* If the data set oscillates, and the length of time between oscillations is uniform, then the data set can be modeled by:
>
> $$f(x) = A * \sin(B(x - C)) + D$$
>
> With:
>
> $$A = \frac{max - min}{2}$$
>
> $$B = \frac{2\pi}{lenght\ of\ the\ oscillation}$$
>
> $$C = \frac{x\ value\ of\ the\ bottom\ of\ an\ oscillation + x\ value\ of\ the\ top\ of\ an\ oscillation}{2}$$
>
> $$D = \frac{max + min}{2}$$

Let's now demonstrate this on a few Examples:

Example 7.4.8: While counting the number of birds in a forest over time, a researcher discovers that on June 2007 there are 125 birds, on December 2007 there are 12 birds, on June 2008 there are 130 birds, on December 2008 there are 9 birds, and on June 2009 there are 119 birds. Create a trigonometric model for the number of birds per month.

Solution to 7.4.8: First of all, we need to verify that the data set is indeed oscillating. In this case, we have a high number of birds in June, and a low number of birds in December, then back to a high number in June. The numbers in June are close to each other, and the numbers in December are close to each other, so a trigonometric model seems like a good model for this situation.

©Brian Gillispie, 2016.

Now, in order to find our trigonometric model, we need to find the values of A, B, C and D for our equation. As stated above, A is the amplitude, or how much the model should vary about the vertical shift. This is found by:

$$A = \frac{130 - 9}{2} = 60.5$$

Next, we need to compute the period. The model oscillates every year, so the period should be 12 months, since we wish to predict the value per month. To make this happen, we need to find:

$$B = \frac{2\pi}{12} = \frac{\pi}{6}$$

To find the phase shift, we need to shift over the function so that $B * (x - C)$ is zero when we are halfway between the max and min. To make this happen, notice the max occurs in month 6, and the min occurs in month 0, assuming we let 0 mean December. Then, this means we need our phase shift to be:

$$C = \frac{6 - 0}{2} = 3$$

Finally, we need to find our vertical shift. The vertical shift can be found by:

$$D = \frac{130 + 9}{2} = 69.5$$

Putting all of this together gives us our final model of:

$$f(x) = 60.5 * \sin\left(\frac{\pi}{6}(x - 3)\right) + 69.5$$

©Brian Gillispie, 2016.

The reader should verify that this model gives the minimum number of birds in December, and the maximum number of birds in June.

Example 7.4.9: While attempting to model at what hour people are more likely to call, a modeler arrives at the following data: At 3 pm, 125 people called, at 6 pm 224 called, at 9 pm 110 called, at midnight 72 called, at 3 am the next day 24 called, at 6 am the next day 5 called, at 9 am the next day 145 called, at noon the next day 151 called, at 3 pm the next day 161 called, and at 6 pm the next day 245 called. Using a trigonometric model where the input is the current hour, create a model that can be used to predict the number of calls that will arrive during any given hour.

Solution to 7.4.9: First, we need to verify that the data set does indeed oscillate. In this case, our peak appears to be at 6 pm, and our low point at 6 am, so the data does oscillate, with equally spaced oscillations, as both are 12 hours apart from each other. Therefore, a trigonometric model will work for this data set.

Start off by finding the amplitude. In this data set, our max is 245, and our min is 5. This gives us an amplitude of:

$$A = \frac{245 - 5}{2} = 120$$

Next, we need to calculate the period. This time we want the data set to model hours, and the model oscillates every day, or every 24 hours. This means our period needs to be:

$$B = \frac{2\pi}{24} = \frac{\pi}{12}$$

Next, find the phase shift. Our min occurs at hour 6, and our max at hour 18, if we let midnight represent hour zero. This gives us a phase shift of:

$$C = \frac{18 + 6}{2} = 12$$

©Brian Gillispie, 2016.

Finally, the vertical shift. In this problem, our vertical shift is:

$$D = \frac{245 + 5}{2} = 125$$

Which, gives us our final model of:

$$f(x) = 120 * \sin\left(\frac{\pi}{12}(x - 12)\right) + 125$$

Before wrapping up this section, we should point out that finding a trigonometric model by this method will not result in the model having no errors. This can be easily seen in the model we just found, as the model claims that at 6 pm on the first day, the center took in 245 calls, not 224 like the center really did. This error results in having the model alternate between the largest value in the data set and the smallest value in the data set. We can usually reduce the by finding the best fitting trigonometric model instead.

Also, the trigonometric model does not handle data sets with multiple peaks. For Example, what if our call center in our Example had peaks at 9 am and 6 pm, but the minimum was still at 6 am. Then we cannot make a trigonometric model like we did here, as it would only account for one of the peaks. There are ways to fix this model to account for the other peak, but they require knowledge of more advanced math techniques, and will not be covered in this book.

©Brian Gillispie, 2016.

7.4 Exercises

For Exercises 1 - 8, state the amplitude, period, phase shift, and vertical shift of the given function.

1. $f(x) = 10 * \sin\left(\frac{\pi}{15}(x - 15)\right) + 25$

2. $f(x) = 63 * \sin(4(x + 20)) - 10$

3. $f(x) = 32 * \sin(4x - 52) + 10$

4. $f(x) = 100 * \sin(\pi x - \pi) - 32$

5. $f(x) = 95.2 * \sin(80x - 120) + 4$

6. $f(x) = -75 * \sin(5x + 60) - 60$

7. $f(x) = -10 * \sin(10x + 10) + 10$

8. $f(x) = -5 * \sin(8x + 120) - 90$

9. For the model that was created in Example 7.4.8, predict how many birds will be present in August 2008.

10. For the model that was created in Example 7.4.9, predict how many calls will arrive at 4 am.

©Brian Gillispie, 2016.

For Exercises 11 - 16, create the appropriate trigonometric model, and use it to answer the desired questions.

11. While attempting to figure out how many ducks are on a pond, a researcher arrives at the following numbers: In December 2008, 6, in March 2009, 45, in June 2009, 156, in September 2009, 98, in December 2009, 12, in March 2010, 52, in June 2010, 153, in September 2010, 76, and in December 2010, 4. Create an appropriate trigonometric model that can be used to predict the number of ducks on the pond in a given month.

12. Create an error table for your model in Exercise 11.

13. Using your model in Exercise 11, predict how many ducks are on the pond in May 2009?

14. Goat populations on an island seem to be following an oscillatory pattern. In December 2009, there were 150 goats, in June 2010 there were 10 goats, in December 2010 there were 152 goats, in June 2011 there were 15 goats, and in December 2011 there were 145 goats. Create an appropriate trigonometric model that can be used to predict the number of goats present on the island on any given month.

15. Create an error table for your model in Exercise 14.

16. Use your model in Exercise 14 to predict how many goats will be present in October 2010.

7.5: Log Scaling

In the process of modeling continuous models, sometimes we will end up with data points that are very spread out, which will make it impossible to visually display the data and show an accurate picture of what is happening. To address this, log scaling of the data points is often used when the data set has both really big and really small values in it. In this Section, we will discuss what log scaling is, and how to use it in a mathematical model.

However, before we begin this Section, be advised that this Section requires that the reader be very familiar with logarithms and exponents and how the two are related[108].

7.5.1: The Idea

To begin our understanding of log scaling, let's look at an Example. Pretend you recorded the following data on energy waves recorded in the ground:

$$6000, 5450, 9000, 1995262, 1122018000, 1995262000000000$$

As you can see, there are a few large values and a few small values. When big numbers arise in a data set along with small numbers[109], it is common to log scale the data so that the data is scaled down to represent powers of 10. This is done by taking the logarithm of the data set that has the small and large numbers. If we take the logarithm of each of these numbers, we end up with:

$$3.78, 3.74, 3.95, 6.30, 9.05, 15.30$$

This new form of the data points has several advantages. The first is that the data values are much smaller and easier to see where they are in relation to each other, especially as it is easy to miss a zero or two or three when looking at the data. The second is this new version of the

---

[108] See Section 6.1 of the book.
[109] Note that if we only had big numbers, we could just rescale using the techniques of Chapter 2. Log scaling is often used when both big and small numbers arise in either the x value or the y value (or both).

©Brian Gillispie, 2016.

data points is easier to graph and display visually as well. If you had attempted to display visually the data before we took the logarithm of the values, you would have either had the data very far apart (and probably not on a single page then), or very clumped together. Neither of these are appealing choices.

Now that we have an idea how to log scale, let's demonstrate log scaling on a few data sets before proceeding.

Example 7.5.1: A modeler has recorded the following data on the number of ants in various colonies.

$$50, 901, 752090, 56000000$$

Log scale the data.

Solution to 7.5.1: To log scale the data, we need to take the logarithm of each of the data points. If we do that, we end up with:

$$1.70, 2.95, 5.88, 7.75$$

Example 7.5.2: A researcher has recorded the following data for the number of bees in various areas:

$$2, 600000, 53000000, 1500$$

Log scale the data.

Solution to 7.5.2: To log scale the data we have to take the logarithm of each of the data points. If we do that, we end up with:

$$0.30, 5.78, 7.72, 3.18$$

©Brian Gillispie, 2016.

Sometimes the data points are scaled again before or after being log scaled. Let's demonstrate that in our next Example.

Example 7.5.3: The following are some recorded intensities of sound, in watts: 0.000001, 0.002, 1, 15, 0.000525. First, log scale the data points, then add 13 to your new points, and then multiply by 10.

Solution to 7.5.3: The easiest way to solve this is to proceed in order. First, take the logarithm of all of the points. If we do that, we end up with:

$$-6, -2.70, 0, 1.18, -3.28$$

Now, add 13 to each of these numbers. Doing that gives us:

$$7, 10.3, 13, 14.18, 10.72$$

Finally, multiply all these points by 10. Doing that gives us:

$$70, 103, 130, 141.8, 107.2$$

Incidentally, what we just did in Example 7.5.3 was we converted from watts to decibels. So now if you wish to know in how many decibels all those sound problems we did in Section 7.3 were, take the logarithm of your answer, then add 13 to it, then multiply by 10.

Sometimes, we need to log scale data points instead of a data set. When you log scale data points, you only take the logarithm of one of the two coordinates in your points. Let's demonstrate that in our next Example.

©Brian Gillispie, 2016.

Example 7.5.4: A researcher has been monitoring the growth of bacteria in a lab, and has noticed that at 2 pm they had 100 cells, at 3 pm they had 500 cells, at 4 pm they had 2500 cells, and at 5 pm they had 12500 cells. Log scale the number of cells for this data.

Solution to 7.5.4: To begin, we need to figure out what the data points are for this problem. Based on the information we have, and the rules from earlier in the book, our data points for this problem will be, with x in hours after noon and y in number of cells:

$$(2, 100), (3, 500), (4, 2500), (5, 12500)$$

Now, we wish to log scale the number of cells. To do that, we need to take the logarithm of the y values of the data points. Do not touch the x values, as those are not being log scaled. If we take the logarithm of the y values, we end up with the following for our rescaled data points:

$$(2, 2), (3, 2.70), (4, 3.40), (5, 4.10)$$

When log scaling data points, it is important to be careful to only log scale the data values that have really big and really small numbers. This will often mean you will log scale only the y value, or only the x value of the data points. Sometimes you will log scale both of them, but it is rare.

7.5.2: Creating a Log Scaled Model

Creating a model that accounts for log scaled data points from data points that are initially not log scaled is done by first scaling one of the data values (either x or y) which have very high and very low numbers[110], then creating the appropriate model for the new points. Let's demonstrate by considering the case where we have 100 cells of bacteria at 2 pm and 12500 cells of bacteria at 5 pm. For that data, our data points are the following:

$$(2, 100), (5, 12500)$$

Now, if we log scale the y values, we end up with these as our new points:

$$(2, 2), (5, 4.10)$$

With these new points, we can create a line for our new, log scaled points. Using the techniques of the earlier Chapters, the appropriate line for these two points is:

$$y = 0.7x + 0.6$$

In this model, x is in hours after noon. However, what are the units of y? To see what units y is in with this new model, we plug in 5. Plugging in 5 into the model returns 4.10. Now, remember that 4.10 was the result we got when we took the logarithm of 12500, which means that 4.10 is the power of 10 needed to get 12500, or, $10^{4.10} \approx 12500$. Therefore, this means that y will always return the power of 10 of the number of cells in this form. If you wish y to still represent the number of cells, that can be done by raising the entire right hand side of the equation to the power of 10, or, by doing the following:

$$y = 10^{0.7x+0.6}$$

---

[110] This will usually be the y values, but there are cases where you may wish to log scale x. Always look which one has really large and really small values, and scale that one.

©Brian Gillispie, 2016.

However, since raising the entire side to the power of 10 defeats the whole purpose of log scaling the data points, this is rarely done in practice. Instead, we have to remember that our y values will be in powers of 10 of the original units.

Now that we have seen how to create a log scaled model, let's demonstrate it with an Example.

Example 7.5.5: A modeler has noticed that in May there were 1000 gnats in an area, and in July there were 100,000,000 gnats in an area. Create a linear log scaled model for this situation.

Solution to 7.5.5: First, we need to first figure out what our data points are. For this problem, our data points are $(5, 1000)$ and $(7, 100,000,000)$. Next, we will log scale the gnats, since the gnats have the really large values. If we take the logarithm of the number of gnats, we get that our new data points are:

$$(5, 3), (7, 8)$$

Now, we create the line which connects these two points. By the techniques of our earlier Chapters, the appropriate line is, with x representing the month of the year, and y representing the power of 10 of the number of gnats present:

$$y = 2.5x - 9.5$$

Next, let's look at our rabbit problem from earlier Sections, and see how we can use log scaling to solve that problem.

Example 7.5.6: In 1859, Thomas Austin released 24 rabbits into the wilds of Australia [11]. In 1865, it is estimated that there were 1430511 rabbits in the wilds of Australia. Create a linear log scaled model, then use your model to predict the number of rabbits present in 1861.

Solution to 7.5.6: For this problem, we have two points, which we will represent by $(0, 24)$ and $(6, 1430511)$, and x is in years after 1859. Now, we will take the logarithm of the number of rabbits, since the number of rabbits is very large in one of our data points. If we log scale the number of rabbits, our points become:

$$(0, 1.38), (6, 6.16)$$

Then, using the techniques of earlier, we end up with the following line:

$$y = 0.7967x + 1.38$$

Now, we need to answer the question. We wish to find the number of rabbits present in 1861. Remembering that x is in years after 1859, we plug in 2 for x, which gives us:

$$y = 0.7967(2) + 1.38 = 2.9734$$

Since y is the power of 10 of the number of rabbits, this means that there are $10^{2.9734} \approx 940$ rabbits present.

Notice the answer for Example 7.5.9 is really close[111] to the answer found via the exponential model in Chapter 6? In general, any time you log scale the y values, the linear model used to connect the log scaled y values will return values very close to the exponential equation used to connect the original, unscaled data points. For this reason, log scaling is often thought of as a way to convert an exponential equation to a linear equation.

---

[111] The real answer is 938 rabbits. The difference is due to the fact we rounded twice while creating the model.

©Brian Gillispie, 2016.

7.5 Exercises

1. Log scale the following data: 0.01, 15, 900, 430900, 19000000

2. Log scale the following data: 0.00000001, 0.0000025, 0.066, 1.02, 150000

3. Log scale the following data: 2, 5, 7.8, 1090, 900000000

4. Log scale the following data: 0.000000000000156, 0.000000000559, 0.0000777, 1.02

5. Log scale the following data, then multiply all answers by 2:

$$4, 8, 500, 1056, 100500$$

6. Log scale the following data, then subtract 10 from each answer:

1000, 2000, 5600, 80500, 950000000000

7. Log scale the following data, then add 13 and multiply by 10 (*See Example 7.5.3*)

$$0.00025, 7, 0.0000000075, 0.00005, 14$$

8. Rework Exercise 7.3.20, but this time once you have found your answer, state your answer in decibels (that is, take the logarithm of your answer, add 13, and then multiply by 10).

9. Due to a major technological breakthrough, a stock of a company that was valued at $0.56 a share in March of 2020 is now valued at $412.67 a share in August 2020. Create a linear log scaled model for this situation.

©Brian Gillispie, 2016.

10. As of 5 pm there were 300 liters of pollutant in a huge pool, but after draining and cleaning, at 10 pm there was only 0.01 liters of pollutant left. Create a linear log scaled model for this situation.

11. As of July 4th there were 200 cells of bacteria in a lab, and as of July 11th there were 150,000,000 cells of bacteria in a lab. Create a linear log scaled model for this situation.

12. An island has 26 rabbits in the year 1976, and 150000 rabbits in the year 1979. Create a linear log scaled model, and use it to predict the number of rabbits that will be present in the year 1981.

13. In the year 2020 a new breed of plant is introduced, which thrives on insects. However, the plant also grows very quickly. If there were 2 plants in the year 2020, and 300 plants in the year 2021, use a linear log scaled model to predict how many of this plant will exist in the year 2025, if nothing is done about it.

14. You have managed to create Lithium 8! If you have 500000 grams of Lithium 8, and one second later you have 250000 grams of Lithium 8, use a linear log scaled model to predict how many grams of Lithium 8 you will have left 6 seconds from now?

15. In the year 2000 a scientist locks up in a vault a million grams of Hollium-166. In the year 3000, the same vault is unlocked, and there are now 561,231 grams of Hollium-166. Create a linear log-scaled model, and use it to predict how many grams of Hollium-166 will be present in the year 4500. *Hint: You may wish to scale x to be out of thousands of years.*

16. A new phone is released that is insanely popular with everyone! If on October 2nd there were only 500 people who had the phone, and on October 7th there were 980,000 people who had the phone, use a linear log scaled model to predict how many people will have the phone on October 11th.

©Brian Gillispie, 2016.

17. A cure for an epidemic has been found! If there are 300,000,000 people who are infected with the disease as of January 29th, and 900,000 people who are infected with the disease as of February 2nd, use a linear log scaled model to predict how many people will be infected as of February 5th.

18. Use your model in Exercise 17 to predict on what day there will be less than one person infected with the disease.

19. A lab has created 500,000,000 grams of Carbon-15, then 44 seconds later they have 1900 grams of Carbon-15. Create a linear log scaled model, then use the model to predict how long until the lab has only 1 gram of Carbon-15 left. *Hint: Be careful what you plug into your log scaled model this time!*

20. Use your model in Exercise 19 to predict how long until the lab has only 0.01 grams of Carbon-15 left.

Chapter 7 Summary

7.1: In this Section, we covered what a quadratic is, as well as how to find the zeros and the maximum or minimum of the quadratic. A quadratic is a polynomial of the form:

$$y = a * x^2 + bx + c$$

The zeros of a quadratic can be found by the quadratic formula, which is:

$$x = \frac{-b \pm \sqrt{b^2 - 4ac}}{2a}$$

And the maximum or minimum of the quadratic exists at the value $x = \frac{-b}{2a}$. This point yields the maximum value if $a < 0$, and this point yields a minimum if $a > 0$.

7.2: In this Section, we discussed how to make a polynomial of degree n if we are given $n - 1$ data points. If we have $n - 1$ data points, then to find a polynomial of degree n, we have to solve a system of n equations with n unknowns to find the coefficients.

©Brian Gillispie, 2016.

7.3: In this Section, we discussed how to use the rules of proportionality to create a model. Specifically, if we know that *y* is directly proportional to *x*, we can model the situation then by the equation:

$$y = kx$$

And, if we know that *y* is inversely proportional to *x*, we can model the situation by the equation:

$$y = \frac{k}{x}$$

7.4: In this Section, we covered how to create a trigonometric model to model data that oscillates, provided the oscillations are equally spaced. If these conditions are met, the model then becomes:

$$f(x) = A * \sin(B(x - C)) + D$$

With:

$$A = \frac{max - min}{2}$$

$$B = \frac{2\pi}{lenght\ of\ the\ oscillation}$$

$$C = \frac{x\ value\ of\ the\ bottom\ of\ an\ oscillation + x\ value\ of\ the\ top\ of\ an\ oscillation}{2}$$

$$D = \frac{max + min}{2}$$

©Brian Gillispie, 2016.

7.5: In this Section, we discussed the idea of log scaling, and how to log scale data points. To log scale data points, take the logarithm of the individual data values, and use that as our new point. This allows us to avoid issues when we have data points that are really small and really large in the same data set, but now the modeler has to remember that the values are in powers of 10, not the original values.

Then, we discussed how you can create a linear log scaled model. To do this, first you have to log scale the appropriate value (or values), then create a linear equation for the new scaled points. This technique is often used to convert from an exponential model to a linear model.

©Brian Gillispie, 2016.

Chapter 8: Model Selection

In the last few Chapters, we have discussed how to create various mathematical models. Now, we are ready to turn our attention to how to decide on which mode to use to model the given data. In this Chapter, we will explore and discuss a few of the various criterion a modeler can use to decide which model to use to model data, when multiple models are valid for the data given.

©Brian Gillispie, 2016.

©Brian Gillispie, 2016.

8.1: Review of Mean and Standard Deviation

In the process of selecting which model to use, we will often be required to compute the mean and standard deviation of various data sets. In this Section, we will review how to calculate the mean and standard deviation. Be advised though that if you have access to a way to calculate the mean and standard deviation via technology, we recommend that you use that instead of attempting to compute these by hand.

8.1.1: Mean

The mean of a data set is another word for the average of the data set, and it is denoted by the symbol $\bar{x}$. To compute the mean, add up all the data values we wish to compute the mean of, then divide by how many of those values there were. This is denoted as follows:

Definition 8.1.1: The mean of a sample, denoted by $\bar{x}$, is calculated as follows:

$$\bar{x} = \frac{x_1 + x_2 + \cdots + x_n}{n}$$

Where:

$x_1, x_2, \ldots, x_n$ are the data points

n is how many data points there are

Let's now demonstrate how to use this definition with a few Examples.

©Brian Gillispie, 2016.

Example 8.1.1: Compute the mean of the following exam scores: 72, 51, 99, 100, 22

Solution to 8.1.1: To compute the mean, we need to add up the five given exam scores, then divide by how many of the exam scores there are. This means we need to compute the following:

$$\bar{x} = \frac{72 + 51 + 99 + 100 + 22}{5}$$

Which, once the calculations are complete, gives us that $\bar{x} = 68.8$. Therefore, the mean of the exam scores is 68.8.

Example 8.1.2: Calculate the mean of the following sales for the week: 1257, 3405, 2995, 1600, 950, 2352, 2475.

Solution to 8.1.2: In this example there are seven data points, and we want to compute the mean of these seven data points. This means need to compute the following:

$$\bar{x} = \frac{1257 + 3405 + 2995 + 1600 + 950 + 2352 + 2475}{7}$$

Which, once the calculations are complete, gives us that $\bar{x} \approx 2147.71$. Therefore, the mean of the sales for that week was about 2147.71.

Oftentimes, we will need to be able to calculate the mean of our x values or the mean of our y values, as these values will be important at times when finding the best fitting model. The next Example demonstrates how to calculate these means.

Example 8.1.3: While observing some rabbits, a researcher notes the following information. On week 1 there are 46 rabbits. On week 2 there are 50 rabbits. On week 3 there are 55 rabbits. On week 4 there are 61 rabbits. Given that $x = week$ and $y = rabbits$, compute the mean of the x values and the mean of the y values.

©Brian Gillispie, 2016.

Solution to 8.1.3: Be careful with this problem, as there are both x and y values, and we want the mean of each individual value. With the given information, our points are the following $(1, 46)$, $(2, 50)$, $(3, 55)$ and $(4, 61)$.

To compute the mean of the x values, notice there are only four x values, and they are 1, 2, 3, and 4. Therefore, the mean of the x values (denoted $\bar{x}$) is as follows:

$$\bar{x} = \frac{1+2+3+4}{4} = 2.5$$

Now, we need the mean of the y values. In this problem, our y values are 46, 50, 55, and 61. Therefore, the mean of the y values (denoted $\bar{y}$) is as follows:

$$\bar{y} = \frac{46+50+55+61}{4} = 53$$

Therefore, the mean of our x values is 2.5, and the mean of our y values is 53.

Typically, we will use the symbol $\bar{x}$ to mean the mean of the x values and $\bar{y}$ to mean the mean of the y values in problems where we calculated both means.

©Brian Gillispie, 2016.

8.1.2: Standard Deviation

The standard deviation of a data set is a measure of how spread out the data is, and is denoted by the symbol s. The standard deviation of a data set is found as follows:

> *Definition 8.1.2:* The **standard deviation** of a sample is computed via the following formula:
>
> $$s = \sqrt{\frac{\sum_{i=1}^{n}(x_i - \bar{x})^2}{n-1}}$$
>
> Where:
>
> $\bar{x} = $ The mean of the sample
>
> $n = $ number of data points in the sample

This one is tough at first glance, so let's decipher it. What the formula is saying is the standard deviation requires you to first compute the difference of each data point from the mean, then, square those differences. Once you have done that, add up all those differences. After that is done, divide by the number of data points minus one, then take the square root of that answer.

Before we proceed, there are a couple of things we need to make you aware of when using this formula. First of all, be very careful when using a calculator. Calculators follow the standard order of operations literally, which means what you get from it may not be what you meant. For example, if you enter -2^2 into most calculators, the calculator is going to interpret that to mean take two, square it, then negate the answer, and will return -4. However, you may have meant for the calculator to compute $-2 * -2$, which has the answer 4, not -4. Make sure to use parenthesis when computing squares of negative numbers, else the answer you get from the calculator may not be accurate.

©Brian Gillispie, 2016.

The other thing to be careful of is, if you decide to use technology to compute standard deviations, make sure you are using the formula to compute the standard deviation of a sample[112] and not the standard deviation of a population. This is because the formula is different depending on whether one is computing the standard deviation of a sample or the standard deviation of a population. A suggestion is to double check the technology method you wish to use against the Examples in the book to make sure the technology method you use is calculating the standard deviation of a sample.

Now, with all of that said, let's demonstrate how to use this formula to calculate the standard deviation by hand[113].

Example 8.1.4: Calculate the standard deviation of the following exam scores: 72, 51, 99, 100, 22.

Solution to 8.1.4: The first step is to find the mean, as we will need it later. The mean of this data set we found back in Example 8.1.1, and it was 68.8. Denote this mean by $\bar{x}$. Now, to compute the standard deviation, it is advised to make a table, as follows, where x_i stands for a data point in our data set. Make the table as follows:

x_i	$x_i - \bar{x}$	$(x_i - \bar{x})^2$
72		
51		
99		
100		
22		
	Sum:	Sum:

The first column is where we will put all the given data points. The second column is where we will calculate the difference between the data point and the mean. The third column is where we will calculate the square of those differences. Also be sure to add a row at the bottom for the sums the second and third columns, as we will need to calculate those as well at some point in time.

---

[112] =stdev.s() in excel as of the writing of this book.
[113] We do advise using technology instead though to compute the standard deviation.

©Brian Gillispie, 2016.

To calculate the second column, we need to take each data point and subtract from it the mean. Some of these answers will be negative, and some will be positive. So, in the row with the 72, we will calculate $72 - 68.8$, which will give us 3.2. In the row with 51, we will calculate $51 - 68.8$ which will give us -17.8. Continuing in this fashion gives us the following:

x_i	$x_i - \bar{x}$	$(x_i - \bar{x})^2$
72	3.2	
51	−17.8	
99	30.2	
100	31.2	
22	−46.8	
	Sum: 0	Sum:

Notice that if we add up the values in the second column, we will get zero. This always happens if all calculations to this point are correct. If you do not get a sum of zero when adding up the second column in this table, this means either your mean is incorrect, or one of your differences is incorrect. Go back and fix the error before proceeding, else all future calculations will be incorrect.

Next, we compute the square of everything in column two. Remember that a negative times a negative is positive, or, in other words, anytime you square something the answer will always be positive. This means that none of the entries in column three can ever be negative, ever. So with this in mind, in the row starting with 72, we compute $3.2 * 3.2$, which is 10.24. Then, in the row starting with 51, we compute $-17.8 * -17.8$, which is 316.84. Continuing in like fashion for all of the rows (except the last) gives us the following table:

x_i	$x_i - \bar{x}$	$(x_i - \bar{x})^2$
72	3.2	10.24
51	−17.8	316.84
99	30.2	912.04
100	31.2	973.44
22	−46.8	2190.24
	Sum: 0	Sum:4365.36

Next, we need to add up all the numbers in the third column. If we do that, we get 4365.36. To finish off our standard deviation calculation, we now need to divide this sum by the

©Brian Gillispie, 2016.

number of data points minus one, then take the square root of that answer. We had 5 data points in this problem, so this means we divide our sum by 4. Then, we need to take the square root of that answer. This means that our standard deviation will end up being:

$$s = \sqrt{\frac{4365.36}{4}} \approx 33.04$$

Therefore, our standard deviation of these exam scores is about 33.04.

Sometimes we need to compute the standard deviation of the x and y values of the data set, as this next Example will demonstrate.

Example 8.1.4: A teacher wishes to create a model to predict final exam performance based on performance up to the exam. The teacher has recorded the following data for a class:

Points Before Final	Percent Grade on Final
280	72
293	88
190	51
280	99
26	2

If x represents the points before the final, and y represents the percent grade on the final exam, calculate the standard deviation of the x values, and the standard deviation of the y values.

Solution to 8.1.4: First, we need to compute the mean of the x value and the mean of the y values. Using the rules from earlier in this Chapter, we find that:

$$\bar{x} = \frac{280 + 293 + 190 + 280 + 26}{5} = 213.8$$

$$\bar{y} = \frac{72 + 88 + 51 + 99 + 2}{5} = 62.4$$

©Brian Gillispie, 2016.

Now that we know the mean of the x and y values, we can now set up our table. Our table for finding the standard deviation of the x values is as follows:

x_i	$x_i - \bar{x}$	$(x_i - \bar{x})^2$
280	66.2	4382.44
293	79.2	6272.64
190	−23.8	566.44
280	66.2	4382.44
26	−187.8	35268.84
	Sum: 0	Sum: 50872.8

Based on this table, the sum of the differences between the x values and the mean, squared is 50872.8. Next, we need to divide by the number of x values minus one, then take the square root. In this problem there are five x values, so we have to divide by four. Plugging all of that into our formula for standard deviation gives us:

$$s_x \approx \sqrt{\frac{50872.8}{4}} \approx 112.77$$

Now, we need to do this again for the y values. Making another table for the y values, and recalling that our $\bar{y} = 62.4$ gives us:

y_i	$y_i - \bar{y}$	$(y_i - \bar{y})^2$
72	9.6	92.16
88	25.6	655.36
51	−11.4	129.96
99	36.6	1339.56
2	−60.4	3648.16
	Sum: 0	Sum: 5865.22

After doing all these calculations, we get that the sum of the differences (squared) is 5865.22. Next, we need to divide by the number of y values minus one, then take the square

©Brian Gillispie, 2016.

root. In this problem there are five y values, so we have to divide by four. Plugging all of that into our formula for standard deviation gives us:

$$s_y \approx \sqrt{\frac{5865.22}{4}} \approx 38.29$$

Therefore, the standard deviation of the x values (denoted s_x) is 112.77, and the standard deviation of the y values (denoted s_y) is 38.29

Sometimes, we will need the variance of a data set instead of the standard deviation. The variance is defined as the square of the standard deviation, or alternatively:

> *Definition 8.1.3:* The **variance** of a data set is the square of the standard deviation, or also noted as:
>
> $$variance = s^2$$
>
> This also means one can compute the variance as follows:
>
> $$s^2 = \frac{\sum_{i=1}^{n}(x_i - \bar{x})^2}{n-1}$$
>
> Variance of a sample is often noted as s^2.

In other words, if the variance of a data set is needed, do the exact same steps as you would do for finding the standard deviation, but leave out the final square root step. For instance, in our last Example, the variance of the x values would be:

$$s_x^2 = 12718.2$$

Note the use of the subscript on the variance symbol to note that it is the variance of the x values.

8.1.3: Using Technology

As the reader can see, calculating the mean and standard deviation by hand can be a long process. Thankfully, Microsoft Excel has built in functions that will allow us to compute both of these values quickly. Let's demonstrate by using the data set from Example 8.1.4. For reference, here is the data set:

Points Before Final	Percent Grade on Final
280	72
293	88
190	51
280	99
26	2

To use Excel for computing the mean and standard deviation, first put the entire data set into Excel. Put all the data points for points before the final in column A, and all the data points for percent grade on the final in column B. If you did it right, your spreadsheet should look like this (the labels in row one are optional, but advised, so that you remember which numbers are from which set later on):

	A	B
1	Points Before Final (x)	Percent Grade on Final (y)
2	280	72
3	293	88
4	190	51
5	280	99
6	26	2

©Brian Gillispie, 2016.

From here, in order to compute the mean of any data set, go to any empty cell (A7 will work well), and enter the following command:

$$= AVERAGE(\text{First cell with data: last cell with data})$$

The equal sign is required, as it tells Excel that you wish it to run a built in function. Without it, the program will think you are entering text, and will do nothing. The colon tells Excel to use all the data in-between those cells on both sides of the colon. Therefore, for our example, to calculate the mean of the data set points before final, we would enter into any empty cell in Excel the following command:

$$= AVERAGE(A2:A6)$$

Once the command has been entered, hit enter. Excel should then calculate the average, and in the cell you typed the command into you will see the number 213.8, which was the average of the x values we previously calculated.

To calculate the standard deviation, also go to an empty cell, and enter the following command[114]:

$$= STDEV.S(\text{First cell with data: last cell with data})$$

Which means, for the data set provided, we would enter:

$$= STDEV.S(A2:A6)$$

Be careful to only have excel compute the standard deviation of the data set, and no extraneous data. A common mistake is to accidentally include your average in the standard

---

[114] If you are using versions of Excel older than 2007, you will need to use a different command, as stdev.s was added to excel in 2007.

©Brian Gillispie, 2016.

deviation calculation. However, since you only want the standard deviation of the data set and nothing else, be sure you only tell it the data set, and nothing else.

And finally, to get the average of the data set percent grade on the final, we would go to an empty cell, and enter the following:

$$= AVERAGE(B2:B6)$$

And, similarly, to get the standard deviation of the data set percent grade on final, we would go to an empty cell and enter:

$$= STDEV.S(B2:B6)$$

If you do this correctly, you should get the same answers we got in Example 8.1.4.

Going forward, we will be using technology to calculate the mean and standard deviation of our data sets. We encourage the reader to do the same, as trying to compute the mean and standard deviation of the upcoming problems by hand is a very long and error prone method, and is not recommended for that reason.

8.1 Exercises

For Exercises 1 - 7, compute a) the mean, b) the variance, and c) the standard deviation

1. Exam scores (in points earned) after an exam: 25, 22, 20, 19, 18, 2

2. Height of students in the same classroom, in inches: 58, 58, 59, 60, 63, 65, 65, 66, 68, 71

3. Points score by the football team, during one season: 0, 7, 7, 13, 15, 16, 17, 19, 21, 22, 28, 34

4. ACT scores from a random sample of 10 test takers: 16, 16, 18, 20, 20, 21, 22, 23, 25, 26

5. Amount spent on textbooks as reported by students: $0, $0, $105.34, $115, $140, $200.50, $250.76, $300.65, $365, $400, $500, $650

6. Commuting times as reported by 5 workers interviewed: 5, 5, 10, 10, 15

7. Survival times of 13 goldfish, bought on the same day and dumped into the same tank. Times are reported in days: 0, 2, 5, 15, 20, 25, 75, 100, 165, 188, 352, 375, 2500

©Brian Gillispie, 2016.

For the following data sets in Exercises 8 - 13, compute a) the mean, b) the variance, and c) the standard deviation for the x values and also the y values

8. Given data points are $(1, 100), (2, 150), (3, 200), (4, 275)$

9. Given data points are $(1, 600), (2, 525), (3, 465), (4, 400)$

10. A researcher is tracking goats on an island. When the researcher arrives, there are 12 goats. One month after the researcher arrived, there are 13 goats present Two months after the researcher arrived, there are 15 goats present. Three months after the researcher arrived, there are 17 goats present.

11. An accident occurs and radioactive material is released into the area around a reactor. When the accident occurs, it is estimated that 1 gram of the substance escaped into the area. 1 year after the accident, it is estimated that 0.65 grams are still present. 2 years after the accident, it is estimated that 0.40 grams are still present. 3 years after the accident, estimates now claim there is 0.30 grams of the substance present.

12. In an attempt to create a model to use high school GPA to predict ACT scores, the following data is recorded:

High School GPA	ACT Score
3.9	31
3.8	29
3.65	24
2.5	22
2.3	19
3.1	26
2.4	27

©Brian Gillispie, 2016.

13. In an attempt to figure out if lower prices cause more sales to occur, the following data is recorded:

Price per Unit	Monthly Sales
$5	5000
$4.50	5500
$4	5750
$3.50	6000
$3	6150
$2.50	5950
$2	6000
$1.75	6750

8.2: Method of Change Comparison

The first method of model selection that we will discussion in this Chapter is the idea of change comparison. The method of model selection by change comparison applies only to models with three or more data points present, and is best applied if we wish to create a linear, exponential, or quadratic model. If other models are desired for the data, other methods should be used.

To use change comparison, first we need to list all of the data points in a table, like we did back in Chapter 3. Then, we need to calculate[115] the Arithmetic Change, the Geometric Change, and the Change of the Arithmetic Change (ΔAC). Then, we will select which model to create by the following rules

---

Definition 8.2.1: The **method of change comparison** creates a model that passes through three or more points via the following rules:

-If the Arithmetic Change of the points is constant (or nearly constant), a linear model is used, with the Arithmetic Change of the points as the slope of the model.
-If the Geometric Change of the points is constant (or nearly constant) an exponential model is used, with the Geometric Change of the points as the base of the exponent.
-If ΔAC is constant, then a quadratic model is used, and the quadratic model is the following:

$$f(x) = \frac{d}{2} * x^2 + \left(b - \frac{d}{2}\right)x + c$$

Where d is the ΔAC and b is the Arithmetic Change between the first two points.

If by any chance multiple changes are constant, you are free to break the tie by any method you wish, as both methods will be equally viable then.

Now, let's demonstrate using this method with a few Examples.

---

[115] Please see Chapter 3 for more on how to calculate these if needed, as this Section assumes you are familiar with how to calculate these already.

©Brian Gillispie, 2016.

Example 8.2.1: A modeler has recorded the following data on cars on the highway at various hours of the day. At 2 pm there were 5600 cars, at 3 pm there were 5700 cars, at 4 pm there were 5800 cars, and at 5 pm there were 5900 cars. Create a model to represent this situation, using the method of change comparison.

Solution to 8.2.1: First, we need to represent our given data as points. If we let x be the time after noon, and y be the number of cars, we end up with the following points, which we put in a table:

Hour	Cars
2	5600
3	5700
4	5800
5	5900

Next, we make a change table[116]. We will first make one for the arithmetic change and the ΔAC, which is as follows:

Hour	Cars	AC	ΔAC
2	5600	NA	NA
3	5700	100	NA
4	5800	100	0
5	5900	100	0

Since the Arithmetic Change is constant[117], we don't need to make any other change tables, and therefore this means we will create a linear model, with the Arithmetic Change as the slope of the model. To create the rest of the linear model, pick any point in our data set[118], and plug that point as well as our slope into the point slope equation for a line. If we do that, we end up with the following linear model:

$$y - 5900 = 100(x - 5)$$

---

[116] See Section 3.5 for a review of how to make them if needed.
[117] Notice we ignore the fact that ΔAC is constant if the Arithmetic Change is also constant.
[118] While any point works, for reasons we will discuss later, we recommend using the **last** point in the data set.

©Brian Gillispie, 2016.

This simplifies to:

$$y = 100x + 5400$$

Which, is our model for the given situation.

Example 8.2.2: A modeler has observed a flock of ducks on a lake, and has recorded the following data. In the month of May, there were 100 ducks, in the month of June, there were 120 ducks, in the month of July there were 144 ducks, and in the month of August there were 173 ducks. Create a model to represent this situation, using the method of change comparison.

Solution to 8.2.2: Again, start out by figuring out what the data points are for this situation. If we let x be the month, and y be the number of ducks, we end up with the following points for our table:

Month	Ducks
5	100
6	120
7	144
8	173

Next, compute the Arithmetic Change and ΔAC in a change table. If we do that, we end up with:

Month	Ducks	AC	ΔAC
5	100	NA	NA
6	120	20	NA
7	144	24	4
8	173	29	5

©Brian Gillispie, 2016.

Neither of these are constant, though the ΔAC is very close and could probably be used in a pinch. Let's check the Geometric Change now with a change table and see what is going on there:

Month	Ducks	GC
5	100	NA
6	120	1.2
7	144	1.2
8	173	1.2014

Looks like the Geometric Change is very close to constant, with the only one that is not exactly 1.2 the very last geometric change we calculated. For that reason, we will select an exponential model, with the value of 1.2 as the base of the exponential model (we called the base a back in our Chapter on exponential models). For our starting value, since we have many data points, we are free to pick one of them to represent our starting value, but whichever one we pick, we will have to shift the model so that an input of 0 returns our starting value for the model. In this case, we will pick the point $(8, 173)$ as the point that has our starting value, which means C is 173 in my model. Therefore, since an input of 0 returns 173, we need to shift the model so that an input of 8 returns 173, which means our final model is:

$$y = 173 * 1.2^{(x-8)}$$

Notice the exponent is x minus the x value of the point that has our starting value. This will always shift the model appropriately so that it returns the appropriate values based on how our data points are set up.

©Brian Gillispie, 2016.

Example 8.2.3: A modeler observing a group of bunnies records the following data. In the year 2020 there were 200 rabbits. In the year 2021, there were 300 rabbits. In the year 2022, there were 500 rabbits, and in the year 2023, there were 800 rabbits. Create a model using the method of change comparison for this situation.

Solution to 8.2.3: Start off by figuring out what our data points are. If we let x be the year in years after 2020, and y be the number of rabbits, our data points are then the following:

Year after 2020	Rabbits
0	200
1	300
2	500
3	800

Calculating the Arithmetic Change and the ΔAC for our points, we end up with the following:

Year after 2020	Rabbits	Arithmetic Change	ΔAC
0	200	NA	NA
1	300	100	NA
2	500	200	100
3	800	300	100

The ΔAC in this data set is constant, so we will use a quadratic model. For the quadratic model, the coefficient in front of x^2 is half of the ΔAC, so it will be 50 in this case. The coefficient in front of the x is the arithmetic change between the first two points, minus half the ΔAC, or $100 - 50$ or 50 in this case. Plugging all of this into our quadratic model, we end up with the following:

$$y = 50x^2 + 50x + c$$

©Brian Gillispie, 2016.

To finish, we need to find c. The easiest way to do that is to plug in one of the points into the equation and solve for it. We will plug in $(3, 800)$ into this model, which, if we do that, we end up with:

$$800 = 50 * 3^2 + 50 * 3 + c$$

Which, if we solve that equation, we end up with:

$$c = 200$$

Therefore, our final model is the following:

$$y = 50x^2 + 50x + 200$$

So far in all of our Examples there has been a clear cut choice as to which model to select, as in all three Examples, one of the changes was so very close to constant it stood out as the obvious choice. But what do you do if none of them stand out like they did in our Examples so far? If that were to happen, the modeler has a few choices as to what to do. They can either use the change that was the closest to being a constant, or they can use another method to select which model to use. In this Section we will use the change that was the closest to being a constant, though which choice you use is up to you to decide.

However, using the change that is closest to being constant will introduce a problem, and that is there is now going to be some error between your model and your data points. Up to this point, any error that was created between the model and the given data points was minimal, but now, if none of the changes are constant, you are guaranteed to create some error between your model and your data points. This can cause a problem if the error is introduced by the last point, as it will throw off future predictions badly.

For Example, assume you have the following data points, and the following changes:

Year	Wolves	GC	AC	ΔAC
2025	60	NA	NA	NA
2026	66	1.1	6	NA
2027	72	1.0909	6	0
2028	290	4.0278	218	212

As you can see, the Arithmetic Change is the closest to being constant, though it has that sudden jump at the end. Still, it is the best choice of the three choices, as none of the others were close to being constant. If we create a line using the Arithmetic Change of 6 as the slope, and the first data point as our point (with x as years after 2000), we end up with the following for our linear model:

$$y = 6x - 60$$

However, if we use this model to predict the number of wolves present in the year 2028, we end up with (and recall that x is years after 2000):

$$y = 6(28) - 60 = 78$$

Which is well off of the real value of 290. This is a problem, because, if a model is off in predicting the last value of the given data set, it is very unlikely to be accurate in predicting the future too. To avoid this, modelers usually use the **last** data point (or last two data points, if two are needed) when creating the model. If we did that, we would find that our model is now:

$$y = 6x + 122$$

While this model will now introduce error when using it to find how many wolves are present in the years 2025, 2026, and 2027, we are not worried about that if our goal is to predict the wolves present in the future. Using the last known data point makes the model more likely to be accurate in making final predictions about future behavior. For this reason, please use the last known data point to create your lines, exponents, and quadratics when working with this method,

©Brian Gillispie, 2016.

as it will make your models much more accurate in predicting the future, even if there is some cost in being able to use it to see what happened in the past.

8.2 Exercises

For all of the Exercises, use the method of change comparison to make the appropriate model. If you are asked to make a prediction as well, be sure to do that too!

1. The data points are $(0, 200), (1, 210), (2, 220), (3, 230)$

2. The data points are $(0, 1000), (1, 950), (2, 900), (3, 850)$

3. The data points are $(0, 1000), (1, 900), (2, 810), (3, 729)$

4. The data points are $(0, 400), (1, 415), (2, 425), (3, 430)$

5. The data points are $(0, 500), (1, 460), (2, 450), (3, 470)$

6. A modeler has been observing sales at a store, and has the following data. On Monday, there were $3000 in sales, on Tuesday there were $3050 in sales, on Wednesday there were $3100 in sales, and on Thursday there were $3150 in sales.

7. Using your model in Exercise 6, predict the amount of sales on Saturday.

8. A modeler has been observing the cost of milk in a town, and has the following data. In the year 2020, milk cost $3, in 2021, milk cost $3.24, in 2022, milk cost $3.50, and in 2023, milk cost $3.78.

9. Using your model in Exercise 8, predict what year the cost of milk will be $5.

©Brian Gillispie, 2016.

10. In a local cave, the bat population has been dropping! In May there were 1000 bats, in June there were 800 bats, in July there were 640 bats, and in August there were 512 bats.

11. Using your model in Exercise 10, predict on which month you have only 100 bats left in the cave.

12. A lake has become polluted! As of May 2nd there were 100 liters of pollutant in the lake, as of May 3rd there was 70 liters of pollutant, as of May 4th there was 49 liters of pollutant, and as of May 5th there was 34.3 liters of pollutant.

13. Using your model in Exercise 12, predict on which day the lake will only have 1 liter of pollutant left.

14. Sales of a popular new CD are dropping. The 6th week after the CD came out, you only sold 1000 copies, the 7th week you only sold 800 copies, the 8th week you only sold 600 copies, the 9th week you only sold 400 copies, and the 10th week you only sold 200 copies.

15. A new phone has been released! The week it is released (week 0) you sell 5000 phones. Week 1 you sell 3500 phones, week 2 you sell 2000 phones, then, due to a new marketing campaign, you sell 50,000 phones on week 3! Afterwards, on week 4, you sell 48,500 phones. Predict how many phones you will sell in week 6, based on this data.

©Brian Gillispie, 2016.

8.3: Method of Least Error

In the last Section, we discussed the idea of change comparison, and how to use it to select which model to use. As we saw, change comparison is based on the idea that if one of the changes is (nearly) constant, we can use that to determine which model to create. However, in many cases, the change between data points is not going to be constant on even close to constant. In those cases, we need another method to determine which model to use. In this Section, we will discuss one such method, which is the method of least error comparison.

The idea behind the method of least error is we will create all the models that we wish to consider, then create an error table and check each model against the given data points, and see which one has the lowest sum of errors to this point. Whichever model has the lowest sum of errors is the model we will select. In the event of a tie, the modeler is free to decide.

Before we demonstrate this technique, we should point out that in the event only two points are given, both the linear and exponential model will have a sum of error of 0, which will force the modeler to use other techniques to determine which model to use. For that reason, we will only use this technique on data sets of three or more points. Let's demonstrate this technique now.

Example 8.3.1: While observing ducks on the local point, you notice that the pond starts out with 20 ducks, then after 1 month it has 25 ducks, then after 2 months it has 31 ducks. Use the method of least error to determine whether to use a linear or an exponential model.

Solution to 8.3.1: Just like before, we have to decide on our data points before proceeding. In this case, our data points are $(0, 20)$, $(1, 25)$, and $(2, 31)$, with x in months since we started observing, and y in number of ducks present.

Next, we need to decide on which model to use. Since we have three data points, we can create both models and see which one best fits the data points. However, our techniques for creating lines and exponents only required two points. Therefore, we will use the first and last data point to create our models[119]. Using the techniques learned earlier for creating linear models, we end up with the following linear model for this situation:

$$y = 5.5x + 20$$

And, using the techniques learned earlier for creating exponential models, we create the following exponential model for this situation:

$$y = 20 * (1.55)^{x/2}$$

Finally, we calculate the error for both models. We do this by making an error table for both models. First, we will consider the linear model.

X	Y (Actual Value)	Calculated Value	Absolute Error
0	20	20	0
1	25	25.5	0.5
2	31	31	0
Sum:			0.5

Notice we don't care about the sign of the error while doing this. Next, the exponential model:

X	Y (Actual Value)	Calculated Value	Absolute Error
0	20	20	0
1	25	24.89	0.11
2	31	31	0
Sum:			0.11

In this Example, the exponential model has the lowest sum of errors, and therefore, based on that, the exponential model is the best model to use. Therefore, our final model to represent this situation is:

---

[119] It is important that you use the same two data points for both models, else the error comparison will be inconsistent between the two models.

©Brian Gillispie, 2016.

$$y = 20 * (1.55)^{x/2}$$

The reader might notice that we did not consider the quadratic model in the last Example, even though if we had created a quadratic model, it would have had the lowest error[120] of all of the cases! The reason the quadratic model was not considered is because creating a quadratic model from data points is very time consuming, plus, in practice, the linear (or exponential) model performs about as well as the quadratic in making short term predictions. For this reason, we will not consider the quadratic model in this Section, and will restrict ourselves to only linear or exponential models.

Next, let's work an Example where we have more than three data points provided.

Example 8.3.2: While observing a flock of birds, you collect the following data. In the month of June, there are 10000 birds, in the month of July there are 11250 birds, in the month of August there are 13000 birds, and in the month of September there are 14200 birds. Create either a linear or an exponential model, using the method of least error.

Solution to 8.3.2: Start off by figuring out what the data points are for this problem. If we let x be months after June, and y be the number of birds, our data points will then be $(0, 10000)$, $(1, 11250)$, $(2, 13000)$, and $(3, 14200)$.

Next, create the models by using the first and last data point. If we do that, we end up with the following linear model and the following exponential model:

$$y = 1400x + 10000$$
$$y = 10000 * 1.42^{(x/3)}$$

Next, create an error table for each model. The error table for the linear model is as follows:

---

[120] Excluding rounding error when creating the coefficients.

©Brian Gillispie, 2016.

X	Y (Actual Value)	Calculated Value	Absolute Error
0	10000	10000	0
1	11000	11400	400
2	13000	12800	200
3	14200	14200	0
Sum:			600

And, the error table for the exponential model is:

X	Y (Actual Value)	Calculated Value	Absolute Error
0	10000	10000	0
1	11000	11240	240
2	13000	12633	367
3	14200	14200	0
Sum:			607

Therefore, since the linear model has the lowest cumulative error, we will use the linear model. That means that for this situation, our model is:

$$y = 1400x + 10000$$

Example 8.3.3: In an attempt to predict the number of customers that need to be served at a local water park, you record the number of customers on five different days. On June 16th, there were 6000 customers in the water park, on June 17th there were 5950 customers, on June 18th there were 6400 customers, on June 19th there were 6100 customers, and on June 20th there were 6600 customers. Use the method of least error to decide on whether to use a linear or an exponential model, and use it to predict the number of customers that will be in the water park on June 23rd.

Solution to 8.3.3: Again, start out by deciding on what the data points are for this situation. If we let x be days after June 16th, and y be the corresponding number of customers, our data points are then $(0, 6000), (1, 5950), (2, 6400), (3, 6100), (4, 6600)$.

Next, we create our linear and exponential models, using the first and last data points. Doing that gives us the following linear and exponential models:

©Brian Gillispie, 2016.

$$y = 0150x + 6000$$
$$y = 6000 * 1.1^{(x/4)}$$

Next, create an error table for this situation. If we do that, the error table for our linear model is:

X	Y (Actual Value)	Calculated Value	Absolute Error
0	6000	6000	0
1	5950	6150	200
2	6400	6300	100
3	6100	6450	350
4	6600	6600	0
Sum:			650

And, our error table for our exponential model is:

X	Y (Actual Value)	Calculated Value	Absolute Error
0	6000	6000	0
1	5950	6145	195
2	6400	6283	117
3	6100	6445	345
4	6600	6600	0
Sum:			657

Since the linear model has the lowest sum of errors, we will use that model to answer our question asked. Recalling that x is days after June 16th, and we wish to predict the number of customers on June 23rd, we compute:

$$y = 150 * 7 + 6000 = 7050$$

Therefore, we conclude that, based on our model, the water park will have about 7050 customers on June 23rd.

©Brian Gillispie, 2016.

While the method of least error works well for distinguishing between when to use a linear and when to use an exponential model when there are very few points, when many data points exist in our set, the method does not work as well, due to our insistence on using the data points at the start and end of the data set. Also, the method of least error performs very poorly if there is a sudden jump in the data set. To see this, try working Example 8.3.3 again, but this time set the number of customers to 200 on June 19th instead. You will notice that now, both methods return a large amount of error.

Another issue with the method of least error is it does not handle oscillatory data well. If your data set alternates between high and low values, both methods will return a high sum of error. To see this, rework Example 8.3.3, but now set June 17th's number of customers to 9000 and June 19th's number of customers to 200. Now both methods will return high error, and will seem to ignore the high value on June 17th and the low value on June 19th in the final model. However, in modeling, those high and low values may tell us something significant that we need to be using in our model[121], and this method is essentially discarding them in the end.

---

[121] For Example, maybe June 17th the high temperature was 104 $F°$, so everyone came to the water park to cool off. Then, on June 19th, there were thunderstorms all day, so very few people came to the water park.

©Brian Gillispie, 2016.

8.3 Exercises

Use the Method of Least Error to create the appropriate linear or exponential model.

1. A zoo has 12,000 people visit in the month of April, 13,500 people visit in the month of May, and 14600 people visit in the month of June.

2. A radioactive material has 100 grams at 12:05 am, 99 grams at 12:09 am, and 98.5 grams at 12:13 am.

3. At age 16, you earn $7.25 an hour. At age 17, you earn $7.55 an hour, and at age 18, you earn $9 an hour.

4. A pack of wolves has 50 wolves in January, 47 wolves in February, 45 wolves in March, and 71 wolves in April.

5. An apple orchard has 6000 apples still on the trees on September 30th, 5850 apples still on the trees on October 1st, 5600 apples still on the trees on October 2nd, and 5400 apples still on the threes on October 3rd.

6. A town has 6500 people in it in the year 2010. One year later (2011) it has 6250 people, and two years later (2012) it has 5800. Predict how many people will be present in the year 2013, using the model of least error.

7. A pollutant has escaped into a lake. Currently there is 100 g of pollutant in the lake, as of 3:50 pm. Massive clean-up efforts begin, and by 4:50 pm there is 90 g of pollutant left. After even more clean-up efforts, by 5:50 pm there is only 77 g of pollutant left. Predict how much pollutant will still be present at 7:50 pm, using the method of least error.

©Brian Gillispie, 2016.

8. On June 8th, a water park had 5000 customers, then on June 9th the water park had 4725 customers, and on June 10th the water park had 5300 customers. Using the model of least error, predict how many people will be in the water park on June 12th.

9. A beekeeper estimates that he sold 700 jars of honey in the year 2008, 800 jars of honey in 2009, and 1000 jars of honey in 2010. Using the model of least error, predict how many jars of honey he will sell in 2012.

10. Back in 1995, gas cost $0.99 a gallon. Then, in 2001, gas cost $1.49 a gallon, and in 2005 gas cost $1.99 a gallon. Using the model of least error, predict how much gas will cost in 2030.

11. A video game that you really want to play retails in February at $59.99. In April, the game goes on sale to $54.99, then in July, the price drops to $49.99. Using the model of least error, predict how much the game will sell for in December, and round your price to $4.99, $9.99, $14.99, $19.99, $24.99, $29.99, $34.99, $39.99, $44.99, or $49.99.

12. An university has 2450 students enrolled in the year 2017, 2500 students enrolled in the year 2018, 2580 students enrolled in the year 2019, and 2700 students enrolled in the year 2020. Using the model of least error, predict how many students will be enrolled in the year 2022.

13. In the year 2020 you are making $15 an hour at your job. In the year 2021, you are making $15.60 an hour. In the year 2022, you are making $16.45 an hour, and in the year 2023, you are making $16.65 an hour. Using the model of least error, predict how much you will be making at that job in the year 2025.

14. While observing some wolves on an island, you record the following data. In 2020 there are 34 wolves, in 2021 there are 44 wolves, in 2022 there are 57 wolves, and in 2023 there are 420 wolves. Using the model of least error, predict how many wolves will be present in the year 2025.

©Brian Gillispie, 2016.

15. A company has recorded the following data for maintenance expenses for the previous years. In 2011, maintenance cost 5 million dollars, in 2012, it was 5.6 million dollars, in 2013, it was 6.1 million dollars, and in 2014, it was 6.2 million dollars. Using the model of least error, predict how much maintenance will cost in 2016 for this company.

16. Repeat Exercise 15, but this time in 2013 maintenance was 16 million dollars.

17. Repeat Exercise 15, but this time in 2014 maintenance was 10 million dollars.

18*. While observing a local pond, you record the following data on the number of ducks on the pond. In the month of February, there were no ducks on the pond. In the month of March, there were 2 ducks. In April, there were 14 ducks, and in May there were 55 ducks. Using the model of least error, predict how many ducks will be on the pond in June? *Hint: You will not be able to use the first and last data points here, as you cannot create an exponential model with a y value of zero. Think of how to adjust the steps used in this Section accordingly so you can still create your models.*

19*. While recording weather data, you record the following information. At 10 pm, the temperature outside is 22 degrees, at 11 pm the temperature outside is 20 degrees, and at midnight the temperature outside is −2 degrees. Using the model of least error, predict the temperature outside at 1 am.

20*. If the temperatures in Exercise 19 are in Fahrenheit, convert them to the Kelvin scale, and redo Exercise 19.

8.4: Method of Lowest Squared Error

In the last Section, we discussed the idea of picking the model based on which model has the least error. In this Section, we are going to modify that method a little, and look at picking the model that has the lowest square of errors present. Modelers often prefer to use the model with the lowest square of errors instead of the lowest total error, due to the fact that models with the lowest error can have a few points which have a very high amount of error, whereas models that have low squared error usually avoid having such points exist in the model.

The technique for finding the model with the lowest squared error is the same as in the last Section, except, now we square our errors when we calculate them on the error table, then calculate the sum of these squared errors (called SSE for short) . Let's demonstrate.

Example 8.4.1: A modeler records the following data for the number of birds in a wildlife reserve. In May there are 12000 birds, in June there are 13500 birds, in July there are 14500 birds, and in August there are 16500 birds. Find the model with the lowest SSE that uses the first and last data points when creating the model.

Solution to 8.4.1: Just like in the last few Sections, start out by defining the points for the model. If we let x be the months after May, and y the number of birds, our data points then will be the following: $(0, 12000)$, $(1, 13500)$, $(2, 14500)$, and $(3, 16500)$.

With the points now defined, create the model using the first and last data point by the methods of the last few Chapters. If we do that, our models are the following:

$$y = 1500x + 12000$$
$$y = 12000 * 1.375^{(x/3)}$$

Next, create the error table for each of these models, and this time, calculate the square of the errors. For the linear model, our error table is as follows:

X	Y (Actual Value)	Calculated Value	Absolute Error	Squared Error
0	12000	12000	0	0
1	13500	13500	0	0
2	14500	15000	500	250000
3	16500	16500	0	0
Sum:			500	250000

And, our error table for our exponential model is as follows:

X	Y (Actual Value)	Calculated Value	Absolute Error	Squared Error
0	12000	12000	0	0
1	13500	13344	166	27556
2	14500	14838	338	114244
3	16500	16500	0	0
Sum:			504	141800

By our calculations, the Exponential model has the lowest SSE, so our model that we will select for this data set will be the exponential model of:

$$y = 12000 * 1.375^{(x/3)}$$

Notice that in Example 8.4.1, the model of the lowest SSE is different than the model with the lowest error! This is because the linear model had only one point with error, but the error on that point was high enough that when we squared the error at that point, it was more than the sum of the squares of all of the errors of the exponential model. Squaring the error at each point puts more emphasis on large errors, and less emphasis on small errors, and as a result, can result in a model with many little errors at each point being selected over a model which has one point with a lot of error.

When using models found by the method of the lowest SSE, modelers often wish to measure how accurately the model is lining up with the data. While the SSE is a decent way to determine this, another measurement is often used for this, which is called R^2. By definition, R^2

is the percent of variance that is explained by the model, and is a number between zero and one. The definition of R^2 is as follows[122].

> *Definition 8.4.1:* **R^2** is defined as one minus the percent of variance the model still has, out of the total variance originally present. If there are only one independent and one dependent quantitative variables present in the model, R^2 is calculated as follows:
>
> $$R^2 = 1 - \frac{SSE_y}{MSE_y}$$
>
> Where:
>
> $SSE_y = $ The Sum of the Squared Errors of the Model
>
> $MSE_y = (n-1) * (s_y)^2$

For the definition, recall that $(s_y)^2$ stands for the variance of the y values, and n is the number of y values. Also, if for some odd reason the model results in worse error than using no model at all, R^2 is reported as 0%, or 0. Now, let's demonstrate using this definition on an Example.

Example 8.4.2: For the model chosen in Example 8.4.2, calculate R^2.

Solution to 8.4.2: Recall first that our model for Example 8.4.2 that minimized the SSE of the data points was:

$$y = 12000 * 1.375^{(x/3)}$$

---

[122] The reader who has studied statistics may wonder where the $n-1$ terms are in the sum of error squared and mean square error terms. If we are dealing with one independent and one dependent variable, all the $n-1$ terms cancel out of the equation. Therefore, we chose to not list them here to simplify the calculations some.

©Brian Gillispie, 2016.

Next, to find R^2, we need to calculate the SSE and the MSE. The SSE we already know from our work of finding the model, and it was 141800. To find the MSE, we need to first calculate the variance of the original y values. Recall that the y values of the data points were 12000, 13500, 14500, and 16500. Using technology, we find that the variance of those four numbers to be 3562500. Therefore, this means the MSE is:

$$MSE = (4-1) * 3562500 = 10687500$$

Therefore, by definition of R^2, this makes the R^2 of our model to be:

$$R^2 = 1 - \frac{141800}{10687500} \approx 0.9867$$

Therefore, for our exponential model which we found in Example 8.4.2, R^2 is about 0.9867, or 98.67%.

Example 8.4.3: A researcher records the following data on the number of rhinos present in a forest. In the year 2020, there were 50 rhinos, in 2021, there were 75, in 2022, there were 76, in 2023, there were 77, and in 2024, there were 78. Find the model which passes through the first and last data points and has the lowest SSE, and calculate R^2 for your model.

Solution to 8.4.3: Start out by defining the points for this situation. We will let x be years after 2020, and y the number of rhinos. This makes our data points then $(0, 50)$, $(1, 75)$, $(2, 76)$, $(3, 77)$, and $(4, 78)$.

Now that we have our points defined, create the linear and exponential models which pass through the first and last data points. If we do that, our models are as follows:

$$y = 7x + 50$$
$$y = 50 * 1.56^{(x/4)}$$

Next, create an error table for the linear model, and calculate the SSE. If we do that, our table is as follows for the linear model:

X	Y (Actual Value)	Calculated Value	Absolute Error	Squared Error
0	50	50	0	0
1	75	57	18	324
2	76	64	12	144
3	77	71	6	36
4	78	78	0	0
Sum:			36	504

And, our table for the exponential model is as follows:

X	Y (Actual Value)	Calculated Value	Absolute Error	Squared Error
0	50	50	0	0
1	75	56	19	361
2	76	62	14	196
3	77	69	8	64
4	78	78	0	0
Sum:			41	621

The linear model has the lowest SSE, so we will select the linear model for this situation. Therefore, our model for this situation is:

$$y = 7x + 50$$

Finally, we need to calculate R^2. To do that, we need to calculate the MSE for the y values of the data points. Recall that the y values are 50, 75, 76, 77, and 78. Using technology to first calculate the variance of the y values, we find that the variance of these y values is 141.7. Therefore, the MSE is:

$$MSE = (5 - 1) * 141.7 = 566.8$$

Which means the R^2 for our model is:

©Brian Gillispie, 2016.

$$R^2 = 1 - \frac{504}{566.8} = 0.1108$$

In the last Example, our R^2 for the model we selected was really low. That was because we did not pick the best linear model to use for this situation. Let's demonstrate.

Example 8.4.4: For the same data set in Example 8.4.3, calculate R^2 for the model $y = 5.8x + 59.6$.

Solution to 8.4.4: To calculate R^2 we need to calculate the MSE of the data points and the SSE of the model. We know from Example 8.4.3 that the MSE of the data points is 566.8, so we only need to calculate the SSE of the model to proceed. To calculate the SSE of the model, we create an error table for the model, and calculate the error and the SSE. If we do that, our table is as follows:

X	Y (Actual Value)	Calculated Value	Absolute Error	Squared Error
0	50	59.6	9.6	92.16
1	75	65.4	9.6	92.16
2	76	71.2	4.8	23.04
3	77	77	0	0
4	78	82.8	4.8	23.04
Sum:			28.8	230.4

Therefore, the R^2 for this model is:

$$R^2 = 1 - \frac{230.4}{566.8} = 0.5935$$

Notice that the R^2 for the model in Example 8.4.4 is better than the R^2 for the model in Example 8.4.3. This is because the model was found via a technique to find the best model with the lowest SSE (and therefore the lowest R^2), and is not just the best model with the lowest SSE that happens to pass through our data points. However, the technique for finding said model is beyond the scope of this book, and for that reason we will only focus on finding the best model that passes through two of our given points (usually the first and last point). Nevertheless, be aware that the model you are finding is usually not going to be the best overall model for the data points, just the best model that meets the given restrictions at this point in time. To find the best model overall that minimizes the SSE for our data points, but is allowed to use any two points when creating the model requires a detailed study of statistics, and I encourage anyone interested in learning more about this to study statistics in more detail.

©Brian Gillispie, 2016.

8.4 Exercises

For Exercises 1 - 6, calculate the MSE_y only

1. Given data points of $(1, 4)$, $(2, 10)$, $(3, 16)$, and $(4, 27)$

2. Given data points of $(0, 15)$, $(1, 39)$, $(2, 13)$

3. Given data points of $(1, 46)$, $(2, 50)$, $(3, 61)$, $(4, 72)$

4. Given data points of $(1, 54)$, $(2, 61)$, $(3, 71)$, $(4, 78)$

5. The local hockey team scored 2 goals on March 12th, 1 goal on March 15th, 4 goals on March 17th, and 0 goals on March 19th.

6. The pizza place you manage sold $1,536 on January 27th, $2,055 on January 28th, $2,133 on January 29th, $1,022 on January 30th, and $3,790 on January 31st.

7. Calculate the model with the lowest SSE for the data points in Exercise 4. Use the first and last data point to create the model.

8. Calculate the model with the lowest SSE for the data points in Exercise 5. Use the first and last data point to create the model.

9. Calculate the model with the lowest SSE for the data points in Exercise 6. Use the first and last data point to create the model.

10. A tube of bacteria has 1000 cells at 2 pm, 1200 cells at 3 pm, 1600 cells at 4 pm and 2200 cells at 5 pm. Create the model which has the lowest SSE for the data points, and uses the first and last data point to create the model.

©Brian Gillispie, 2016.

Exercise 11 - 16 are the same as Exercises 1 - 6 from Section 8.3. For all of them, calculate the model with the lowest SSE, using the first and last data points to create your model.

11. A zoo has 12,000 people visit in the month of April, 13,500 people visit in the month of May, and 14600 people visit in the month of June.

12. A radioactive material has 100 grams at 12:05 am, 99 grams at 12:09 am, and 98.5 grams at 12:13 am.

13. At age 16, you earn $7.25 an hour. At age 17, you earn $7.55 an hour, and at age 18, you earn $9 an hour.

14. A pack of wolves has 50 wolves in January, 47 wolves in February, 45 wolves in March, and 71 wolves in April.

15. An apple orchard has 6000 apples still on the trees on September 30th, 5850 apples still on the trees on October 1st, 5600 apples still on the trees on October 2nd, and 5400 apples still on the threes on October 3rd.

16. A town has 6500 people in it in the year 2010. One year later (2011) it has 6250 people, and two years later (2012) it has 5800. .

17. For your model in Exercise 16, predict how many people will be in the town in the year 2013?

18. Calculate R^2 for your model in Exercise 16.

©Brian Gillispie, 2016.

19. A hotel sold 22 rooms on August 1st, 24 rooms on August 2nd, 25 rooms on August 3rd, and 33 rooms on August 4th. Find the model with the lowest SSE which uses the first and last data points to create the model, and calculate R^2 for your model.

20. If the hotel sells out when it sells 40 rooms, predict on what day that will occur, using your model in Exercise 19.

21. On May 17th the river is 12 feet high, on May 18th the river is 12.4 feet high, on May 19th the river is 11.9 feet high, on May 20th the river is 12.2 feet high, and on May 21st the river is 14 feet high. Find the model with the lowest SSE which uses the first and last data points to create the model, and calculate R^2 for your model.

22. Use your model in Exercise 21 to predict how high the river will be on May 23rd.

23. In the year 2016, your business has 4 million dollars in expenses, in 2017 you had 3.75 million in expenses, in 2018 you had 4.4 million in expenses, in 2019 you had 5.2 million in expenses, and in 2020 you had 4.8 million in expenses. Find the model with the lowest SSE which uses the first and last data points to create the model, and calculate R^2 for your model.

24. Use your model in Exercise 23 to predict how much in expenses your business will have in the year 2021.

8.5: Modeling Categorical Data

So far in this Chapter, we have seen various ways to select which model to use when the data is quantitative, or can be converted to quantitative. In this Section, we are going to discuss cases where that is not possible, or in other words, how to model data when one of the data sets is categorical, and there is no logical way to assign numbers to the categorical data.

To begin our discussion of modeling categorical data, let's look at an example. Pretend you wish to see if the color of a new candy determines which candy sells. You place an equal number of each color of candy out on display, then record the amount sold by the end of the week. The following is your data set:

Color	Number Sold
Blue	2,979
Red	1,401
Yellow	1,000
Black	4,052
White	2,221
Purple	4,985
Orange	1,452

Notice that for the variable color, none of the data entries are numbers this time, nor can a logical number assignment be found. Still, it is possible to create a model for the situation despite that. To make a model, start by assigning categories to each of the data. Which category you lump the data to is your choice. For instance, we could break this data into dark colors and light colors. Or we could categorize by colors that start with B and R and colors that do not. The choice of which categories to break the model into is up to the modeler.

Unfortunately we will not be able to show you a way to decide on which categories to break the model into, as that is a trial and error approach. Instead, what we will show is how to determine how effective the choice of the categories for the model were, once they have been decided on. This is determined by the calculation of R^2 of the categorization [14]. However, our calculation of R^2 is a little different due to one of the data sets being categorical, and is defined as follows.

©Brian Gillispie, 2016.

> *Definition 8.5.1:* If the independent variable (x) is categorical and the dependent variable (y) is quantitative, then the R^2 for a categorization is as follows:
>
> $$R^2 = 1 - \frac{SSE}{MSE}$$
>
> Where:
>
> $$MSE = \frac{(n-1) * (s_y)^2}{N-1}$$
>
> $$SSE = \frac{(n_1 - 1) * (s_{y1})^2 + (n_2 - 1) * (s_{y2})^2 + \cdots + (n_m - 1) * (s_{ym})^2}{N-1}$$
>
> n_i = number of data points in the ith categorization
>
> $(s_{yi})^2$ = variance in the ith categorization
>
> N = Total data points in the data set

The easy way to remember the formula for R^2 for categorical data is R^2 is the percent of variance that is explained, after the categorization has been applied. In other words, R^2 is one minus the variance after the categorization divided by the variance before the categorization. If the categorization is good, then there will be little variance within each categorization, and the SSE will be tiny. If not, then there will still be a high variance in each categorization, and the SSE will be large.

Before proceeding, note that how you round the calculations for the variance does matter in what you will get for your final answer for R^2. Rounding differences in how different computers compute the variance can cause a difference in your final answer for R^2 of 1% or so. Please keep this mind when you work out the examples coming up, as it is possible to do the exact same steps we do, and get an R^2 of 79.14% instead of say 79.25%. This difference is minor, and unavoidable, due to the different ways computers round digits. .

Now, let's demonstrate how to calculate R^2 for categorical data.

Example 8.5.1: Which airline has the most delays? In an attempt to answer this, one airport records the number of flights that are on time (out of 10) for a few different airlines. The data is displayed below:

Airline	Number on Time
Chicago Airlines	3
Alaska Airlines	1
Phoenix Airlines	8
Dallas Airlines	7
Seattle Airlines	2
Houston Airlines	6

In an attempt to categorize this data, the company wishes to categorize this data by the towns Chicago, Alaska and Seattle, and Phoenix, Dallas, Houston. Compute R^2 for the categorization.

Solution to 8.5.1: Start out by computing the variance of the original data set before categorization. Using technology, we find that:

$$(s_y)^2 = 8.3$$

Which gives us that our mean square error[123], or MSE is:

$$MSE = \frac{(6-1) * 8.3}{(6-1)} = 8.3$$

Next, put the data sets into their categorizations. Since the tables are small, we will not list them in table form this time. The first categorization is Chicago, Alaska, and Seattle, which had 3, 1, and 2 flights on time respectively. The variance of this data group is:

---

[123] Note that the formula for MSE has an $(n-1)$ divided by an $(N-1)$. For categorical data, those two are always the same, and cancel each other. Therefore, the MSE is always the variance of the original data set, when using our definition.

©Brian Gillispie, 2016.

$$(s_{y1})^2 = 1$$

Similarly, the second categorization includes Phoenix, Dallas, and Houston, with 8, 7, and 6 flights that are on time respectively. Computing the variance for this data group gives us:

$$(s_{y2})^2 = 1$$

Plug these numbers into our formula for SSE, and recall that each data set had 3 data points, so, since each variance has to be multiplied by the number of points in its respective categorization minus one, we will be multiplying each variance by 2. With that in mind, and remembering that there are 6 total data points (so $N - 1$ is 5):

$$SSE = \frac{2*1 + 2*1}{5} = 0.8$$

Plugging this into our formula for R^2 gives us:

$$R^2 = 1 - \frac{0.8}{8.3} \approx .9036$$

This means that our R^2 is about 90.36%, and this categorization explained about 90.36% of the variance between the original data points.

Example 8.5.2: A candy company wishes to determine which color candies are the most popular. To determine this, they make the same amount of each candy, then put them on display and see what the customers buy. One week later they tally up the sales for each color candy. The total candy sales for each color are listed in the following table:

Color	Number Sold
Blue	2,979
Red	1,401
Yellow	1,000
Black	4,052
White	2,221
Purple	4,985
Orange	1,452

However, they wish to see if there is some pattern to which colors sell the most. In an attempt to do this, they break the table into two tables, one with all the bright colors (White, Orange, and Red, Yellow) and one with all dark colors (Blue, Purple, Black). Compute the R^2 of this categorization.

Solution to 8.5.2: To solve this problem, start out by computing the variance of the original data set. Using technology, we find that the variance of the original data set, which we will call $(s_y)^2$, is, rounded to the nearest whole number[124]:

$$(s_y)^2 = 2232931$$

Which, since the MSE is the same as the variance of the entire data set, this gives us that our MSE is:

$$MSE = 223931$$

---

[124] While rounding will introduce some roundoff error in the final answer, when the variance is in the millions, rounding the nearest whole number will usually affect the final answer by a very tiny amount

©Brian Gillispie, 2016.

To compute SSE, we need to compute the variance of each of the data tables under the new categorization. Start out by creating the new tables. We will make one table for the bright colors, and one for the dark colors. The table for the dark colors is as follows:

Color	Number Sold
Blue	2,979
Black	4,052
Purple	4,985

And the table for the light colors is as follows:

Color	Number Sold
Red	1,401
Yellow	1,000
White	2,221
Orange	1,452

Next, compute the variance for each table. Note that each table has a different number of data points, so we will need to take the variance times the number of data points (minus one) in its respective table before adding them together. We will call the variance of the dark colored table $(s_{y1})^2$, and the variance of the light color table $(s_{y2})^2$. Using technology to compute the two variances finds that the variances are, again rounded to the nearest whole number:

$$(s_{y1})^2 = 1007642$$
$$(s_{y2})^2 = 260192$$

Plug these numbers into our SSE formula, and remembering that our first categorization had 3 data points (so $n - 1$ is 2 for this table) and our second categorization had 4 data points (so $n - 1$ is 3 for this table) gives us:

$$SSE = \frac{2 * 1007642 + 3 * 260192}{6} \approx 465977$$

©Brian Gillispie, 2016.

MSE tells us that the original data set had a variance of a little over two million, and the SSE tells us that the categorized data has a variance within each data table of about half a million. Since R^2 is one minus the percent of variance that is still not explained divided by the original variance, that means that:

$$R^2 = 1 - \frac{465976}{2232931} \approx .7779$$

This means that our R^2 is about 77.79%, and our categorization explained about 77.79% of the variance in the sales.

Example 8.5.3: A scientist is interested in determining what causes height to vary as much as it does. He collects data from 9 women, and records the data set[125]. In an attempt to categorize this data, the scientist decides to categorize the data by first letter of the alphabet. Compute R^2 for this data set.

Name	Height in inches
Amy	66
Anne	63
Angela	71
Sarah	72
Susan	59
Sylvie	69
Jennifer	61
Julia	65
Jasmine	72

Solution to 8.5.3: Start out by computing the variance of the original data set. Using technology, we find that the variance of the original data set is:

$$(s_y)^2 = 23.52\overline{7}$$

Therefore, the MSE is:

---

[125] This data set is fictional. In reality, no data set would be published with the real names still included.

©Brian Gillispie, 2016.

$$MSE = 23.52\overline{7}$$

Next, compute the variance for each categorization. This time we have three categorizations, one for the A names, one for the S names, and one for the J names. The A names are Amy, Anne, and Angela, and the heights are 66, 63, and 71 respectively. Computing the variance for that data set gives us:

$$(s_{y1})^2 = 16.\overline{3}$$

For the second categorization, we have Sarah, Susan, and Sylvie, with heights 72, 59, 69, respectively. The variance for this data set is:

$$(s_{y2})^2 = 46.\overline{3}$$

Finally, the third categorization has Jennifer, Julia and Jasmine, with heights 61, 65 and 72 respectively. The variance for this data set is:

$$(s_{y3})^2 = 31$$

Each categorization has three data points, so $n - 1$ is 2 in all cases. Plugging all of this into the formula for SSE, and recalling that there are 9 total data points (so $N - 1$ is 8) gives us:

$$SSE = \frac{2 * 16.\overline{3} + 2 * 46.\overline{3} + 2 * 31}{8} \approx 23.41\overline{6}$$

Plugging in the calculations for MSE and SSE into our formula for R^2 gives us:

$$R^2 = 1 - \frac{23.41\overline{6}}{23.52\overline{7}} \approx 0.0047$$

©Brian Gillispie, 2016.

This means that with these categorizations, R^2 is 0.47%, and this categorization explains only 0.47% of the variance in the data set. This tells us that categorizing women by first letter explains little about why some are tall and some are short, and there is some other factor at work here.

The last Example shows that the calculation for R^2 can be extended to more than two categories. Just remember to always multiply the variance by the number of data points in the categorization minus one, then after adding up all of those numbers, divide the final answer by the total number of data points minus one. A common mistake in this calculation is using the wrong value for n, or using n instead of the number of data points minus one.

©Brian Gillispie, 2016.

8.5 Exercises

1. The following is a list of the number of students in the library per day of the week. Sunday, 925, Monday, 522, Tuesday, 605, Wednesday 669, Thursday 1054, Friday 102, Saturday 95. Categorize the data into Monday - Thursday and Friday - Sunday, and compute R^2 for this categorization.

2. The data set in Exercise 1 had peaks on Sunday and Thursday. Redo Exercise 1, but this time put Sunday and Thursday in their own category, and all other days into their own category. What is R^2 for this categorization?

3. In an effort to determine why some people are short and some are tall, a scientist records the following data. Brett 65, Amy 62, Tina 64, Jake 71, John 70, Sylvie 70, Tony 74, Kylie 60, Richard 64, Ben 76, Laura 72, Lisa 70. If Amy, Jake, Sylvie, Kylie, and Ben all played an instrument as a kid, categorize the data set by those who played an instrument and those who did not. What is R^2 for this categorization?

4. In Exercise 3, Brett, Amy and Tina were all from one family, Jake, John, Sylvie and Tony from another, Kylie and Richard from another, and Ben, Laura and Lisa from another. Redo Exercise 3, categorizing by kids from the same family. What is R^2 for this categorization?

5. The following are lists of test scores on a hard final exam: 98, 95, 92, 90, 88, 88, 87, 85, 72, 55, 32, 31, 22, 19, 18, 5. If the first eight test scores were from students who had done the homework and the last eight test scores were from students who had not done the homework, categorize the data set into students who did and who did not do the homework, and find R^2 for this categorization.

©Brian Gillispie, 2016.

6. An employee comes up for his annual review, and in the process of figuring out the review, the boss figures out how many days the employee has missed per month. The following are the number of days the employee missed by month: January, 2. February, 1. March, 0. April, 1. May, 7. June, 8. July, 10. August, 9. September, 4. October, 0. November, 1. December, 1. Categorize the missed days by the season the missed day occurred in (you decide which season each month is in, but justify the choice!), then compute R^2 for your categorization.

7. If the same employee had been seriously sick in the months of May - September, redo Exercise 7, but this time categorize by months when the employee was seriously sick and months when they were not.

8. A scientist logs his hours spent on the computer by the month, and the results are as follows for each of the months in the year 2016: January, 140. February, 98. March, 122. April, 199. May, 98. June, 21. July, 2. August, 2. September, 9. October, 10. November, 7. December, 2. If the scientist's graduate thesis was due at the end of May 2016, categorize by months before the thesis was due, and months after the thesis was due, and calculate R^2 for your categorization.

9. In an attempt to figure out why employees skip work, a manager records the number of employees absent of a workforce of 50 employees, based on the day of the week. On Monday, 15 were absent. On Tuesday, 2 were absent. On Wednesday, 4 were absent. On Thursday, 6 were absent. On Friday, 22 were absent. Categorize these results into any categories you think are significant. Why did you pick those categories? Calculate R^2 for your categories as well.

©Brian Gillispie, 2016.

Chapter 8 Summary

8.1: In this Section, we reviewed what the mean, standard deviation, and variance are. The mean we defined by:

$$\bar{x} = \frac{x_1 + x_2 + \cdots + x_n}{n}$$

The standard deviation we defined by:

$$s = \sqrt{\frac{\sum_{i=1}^{n}(x_i - \bar{x})^2}{n-1}}$$

And, the variance was defined as the square of the standard deviation, denoted s^2.

8.2: In this Section, we discussed the method of change comparison for deciding which model to use. If the Arithmetic Change is constant, we used a linear model to model the data set, if the Geometric Change is constant, we used an exponential model to model the data set, and if the Change of the Arithmetic Change (ΔAC) is constant, we used a quadratic to model the data set.

8.3: In this Section, we discussed the method of least error to decide whether to use linear or exponential models to model the data sets. To determine which model to use, we first create the model using the first and last data points in the set, then create an error table for each model. The model with the lowest sum of errors is the model we select. In the event of a tie, the modeler is free to choose either model to use.

©Brian Gillispie, 2016.

8.4: In this Section, we discussed how to use the method of SSE, or Sum of Squared Errors, to determine which model to select. To use the method of SSE, first we make each model, using the first and last data points in the set, then we calculate the error and the square of each error, then sum the squares of the errors. The model with the lowest sum of squared errors is the model we select.

In addition, we also discussed how to calculate the value R^2 to determine how well a model fit a given data set. For a given data set, R^2 is:

$$R^2 = 1 - \frac{SSE_y}{MSE_y}$$

Where:

$SSE_y = \Sigma(\tilde{y} - y)^2$, or the sum of the squared errors in the data set.

$$MSE_y = \sum (y - \bar{y})^2$$

©Brian Gillispie, 2016.

8.5: In this Section, we discussed how to use R^2 to determine how efficient a categorization was. The R^2 for a categorization is:

$$R^2 = 1 - \frac{SSE}{MSE}$$

Where:

$$MSE = \frac{(n-1) * (s_y)^2}{N-1}$$

And:

$$SSE = \frac{(n_1 - 1) * (s_{y1})^2 + (n_2 - 1) * (s_{y2})^2 + \cdots + (n_m - 1) * (s_{ym})^2}{N-1}$$

With $(s_{yi})^2$ meaning the variance of the ith categorization, and n_i the number of data points in the ith categorization.

©Brian Gillispie, 2016.

Chapter 9: Other Topics

So far in this book we have discusses many ways to model a given situation. In this Chapter, we are going to discuss a few topics that are worthwhile to know when modeling. These are the topics of how to find a zero of a model, how to set up a domain restricted model, how to set up a piecewise model, and how to work with a parameterized model.

©Brian Gillispie, 2016.

9.1: Zero Finding

While working with a mathematical model, it is common to need to be able to find the zero of the model. Unfortunately, this can be a challenging task for certain types of models, as some models have no algebraic method that can be used to find a zero of that model. In this Section, we will discuss a couple of methods that can still be used to find a zero of the model in these circumstances.

9.1.1: Bisection Method

The bisection method is one of the more popular methods used to find a zero of a function. It is based on what is known as the Intermediate Value Theorem, which we will define now:

> *Definition 9.1.1:* The **Intermediate Value Theorem (IVT)** states that if $f(x)$ is a continuous function on the interval $[a, b]$, and if $f(a) < 0$ and $f(b) > 0$, then for some number c in the interval $[a, b]$, $f(c) = 0$

In other words, what the Intermediate Value Theorem states is provided we know that our function is continuous[126], and provided we know an input of the function where the function output is negative, and an input of the function where the function output is positive, then somewhere between them, the function must return zero, as otherwise, the function would cease to be continuous.

The bisection method uses this fact in the following way. Assume we know two numbers, a_n and b_n, and $f(a_n) < 0$ and $f(b_n) > 0$. Then, compute a new number $c_n = \frac{a_n + b_n}{2}$, and calculate $f(c_n)$. If $f(c_n) < 0$, then let $a_{n+1} = c_n$ and $b_{n+1} = b_n$. If not, then let $a_{n+1} = a_n$ and $b_{n+1} = c_n$. This will set us up so that our next values of a and b return outputs that are opposite

---

[126] All of our functions in Chapters 5 - 7 are continuous in their domain.

©Brian Gillispie, 2016.

in sign[127]. Then, we repeat the method as often as we wish, until we feel that we have a good estimate for the zero of the function. Let's demonstrate this method on a function we know the zero of first, so we can see it in action.

Example 9.1.1: Given the equation $y = -16t^2 + 64t + 32$, find a zero of the equation via the bisection method.

Solution to 9.1.1: Notice this is the same equation[128] we did in Example 8.1.4, where we found the zero of the equation was about 4.45. This time, we will find the zero by the bisection method. To begin, we need to find two inputs in our domain, one which returns a negative number and one which returns a positive number. Plugging in random numbers for a while, we see that if our input is 4, we get that y is 32, and if our input is 5, we get that y is -48. Therefore, we will start our bisection method with $a_0 = 4$ and $b_0 = 5$. From here, we make a table, as follows:

n	a_n	b_n	$f(a_n)$	$f(b_n)$	$c_n = \dfrac{a_n + b_n}{2}$	$f(c_n)$
0	4	5	32	-48	4.5	-4

We calculate c_0 by averaging the two numbers (a_0 and b_0), and we get that c_0 is 4.5, and then we plug in our value for c_0 into the equation to get that $f(c_0)$ is -4. Then, we select the values for a_1 and b_1 in such a way that one of them returns a negative number when we plug it in, and one of them returns a positive number when we plug it in. Looking at the chart, we see that the only way to do that is to pick 4 and 4.5 for our new inputs, so we let a_1 be 4 and b_1 be 4.5. This updates the chart as follows:

n	a_n	b_n	$f(a_n)$	$f(b_n)$	$c_n = \dfrac{a_n + b_n}{2}$	$f(c_n)$
0	4	5	32	-48	4.5	-4
1	4	4.5	32	-4	4.25	15

---

[127] This is **very** critical. If you have $f(a_n) > 0$ and $f(b_n) < 0$, the steps reverse! You must make sure that you always have it so that $f(a_{n+1})$ and $f(b_{n+1})$ are opposite in signs, else this method will fail.
[128] While the method is defined for a function and we have an equation instead, remember that y and f(x) are interchangeable. We will use that rule regularly throughout this Section.

©Brian Gillispie, 2016.

Again, get c_1 by taking the average of a_1 and b_1, then compute the value of the function when c_1 is plugged in. Now, select the values for a_2 and b_2, remembering that one of them has to return a negative output and one of them has to return a positive output. The only way to do that is to let a_2 be 4.25, and b_2 be 4.5. This makes our chart now the following:

n	a_n	b_n	$f(a_n)$	$f(b_n)$	$c_n = \dfrac{a_n + b_n}{2}$	$f(c_n)$
0	4	5	32	−48	4.5	−4
1	4	4.5	32	−4	4.25	15
2	4.25	4.5	15	−4	4.375	5.75

From here, keep up the process until you feel you have a good estimate of where the zero is. For reference, below we have the table taken out a few more steps.

n	a_n	b_n	$f(a_n)$	$f(b_n)$	$c_n = \dfrac{a_n + b_n}{2}$	$f(c_n)$
0	4	5	32	−48	4.5	−4
1	4	4.5	32	−4	4.25	15
2	4.25	4.5	15	−4	4.375	5.75
3	4.375	4.5	5.75	−4	4.4375	0.9375
4	4.4375	4.5	0.9375	−4	4.46875	−1.515625
5	4.4375	4.46875	0.9375	-1.515625	4.453125	−0.28515625
6	4.4375	4.453125	0.9375	−0.28515625	NA	NA

We stopped at step 6, so we put NA in the last two rows. Based on our work, the zero of this equation appears to occur at 4.4, as those are the digits that are the same in both a_6 and b_6 when we stopped.

Now, let's work one where we don't know the zero from any of our past work.

©Brian Gillispie, 2016.

Example 9.1.2: Find a zero of the equation $y = x^3 - 2x^2 + 2x + 3$

Solution to 9.1.2: Like before, we first have to find two inputs, one which returns a negative output, and one which returns a positive output. Playing around with some inputs, we eventually discover that an input of -1 returns -2 and an input of 0 returns 3. Therefore, we set a_0 to be -1 and b_0 to be 3 for the bisection method, and set up our table as follows:

n	a_n	b_n	$f(a_n)$	$f(b_n)$	$c_n = \dfrac{a_n + b_n}{2}$	$f(c_n)$
0	-1	0	-2	3	-0.5	1.375

For the first iteration of the bisection method, our c_0 is -0.5, as $\dfrac{-1+0}{2} = -0.5$. Plugging that into our equation, we get that an input of -0.5 returns 1.375. Therefore, since we need to pick a_1 and b_1 in such a way that one of them returns a negative number when we plug it in, and one of them returns a positive number when we plug it in, we pick a_1 to be -1 and b_1 to be -0.5. This makes our chart now the following:

n	a_n	b_n	$f(a_n)$	$f(b_n)$	$c_n = \dfrac{a_n + b_n}{2}$	$f(c_n)$
0	-1	0	-2	3	-0.5	1.375
1	-1	-0.5	-2	1.375	-0.75	-0.046875

Again, get c_1 by taking the average of a_1 and b_1, then compute the value of the equation when c_1 is plugged in. Now, select the values for a_2 and b_2, remembering that one of them has to return a negative output and one of them has to return a positive output. The only way to do that is to let a_2 be -0.75, and b_2 be -0.5. This makes our chart now the following:

n	a_n	b_n	$f(a_n)$	$f(b_n)$	$c_n = \dfrac{a_n + b_n}{2}$	$f(c_n)$
0	-1	0	-2	3	-0.5	1.375
1	-1	-0.5	-2	1.375	-0.75	-0.046875
2	-0.75	-0.5	-0.046875	1.375	-0.625	≈ 0.73

©Brian Gillispie, 2016.

From here, we continue in the same way as before, stopping when we have a good idea of where the zero is. For reference, below we have the table taken out a few more steps.

n	a_n	b_n	$f(a_n)$	$f(b_n)$	$c_n = \dfrac{a_n + b_n}{2}$	$f(c_n)$
0	-1	0	-2	3	-0.5	1.375
1	-1	-0.5	-2	1.375	-0.75	-0.046875
2	-0.75	-0.5	-0.046875	1.375	-0.625	≈ 0.73
3	-0.75	-0.625	-0.046875	≈ 0.73	-0.6875	≈ 0.35
4	-0.75	-0.6875	-0.046875	≈ 0.35	-0.71875	≈ 0.15
5	-0.75	-0.71875	-0.046875	≈ 0.15	-0.734375	≈ 0.06
6	-0.75	-0.734375	-0.046875	≈ 0.06	NA	NA

Like before, we stopped when n was 6, so we put NA for the values we didn't compute. Therefore, based on our work so far, we conclude that $x^3 - 2x^2 + 2x + 3 = 0$ when $x \approx -0.7$.

While the bisection method does work and eventually will find you the zero of the function, provided all the conditions are met, there are a few problems to be aware of when using it. First, the bisection method is slow [13]. In bot of our Examples, we were only able to find our zero to the nearest tenth after 6 steps of the method. To find our zero with more accuracy, many, many more steps are needed. Second, the bisection method only returns one of the zeros in the interval that you start in. If there are multiple zeros in the starting interval, the method will find only one of them [13], and which one the method finds depends both on what interval you start on, and on the function the method is being applied to.

©Brian Gillispie, 2016.

9.1.2: Secant Method

The Secant Method is another method that is used to find the zero of an equation or function. The Secant Method is based on the idea that if we have two inputs to our function, x_n and x_{n-1} that are close[129] to our desired zero, then, we draw a line between the points $(x_n, f(x_n))$ and $(x_{n-1}, f(x_{n-1}))$, and find where that line is zero. Next, use the input that made that line zero, and the previous input x_n and repeat as needed.

Because the method repeats and does not require a selection process like the bisection method did, it can be defined by a recursive sequence. The recursive sequence for the secant method is as follows:

> *Definition 9.1.2:* Given two initial inputs x_0 and x_1 which are 'close' to the zero of the function $f(x)$, the inputs to the **Secant Method** can be defined by the following recursive relationship.
>
> $$x_n = x_{n-1} - f(x_{n-1}) * \frac{x_{n-1} - x_{n-2}}{f(x_{n-1}) - f(x_{n-2})}$$

The recursive sequence used for the Secant Method is a second order recursive relationship, because it depends on two of the previous values, x_{n-1} and x_{n-2}. Also, remember that $f(something)$ means to plug that into the function or equation and evaluate it.

Now, let's demonstrate the Secant Method with our equation that we used in Example 9.1.1.

Example 9.1.3: Find a zero of the equation $y = -16t^2 + 64t + 32$ via the Secant Method

Solution to 9.1.3: To begin, we need to find our first two values of the sequence x_0 and x_1. Unlike with the Bisection Method, these two do not have to return an output where one output is negative and one output is positive. Instead, the points need to be 'close' to the zero. However,

---

[129] How close varies by too many factors to be mentioned. All you can do is use your judgment.

©Brian Gillispie, 2016.

we will go ahead and start with the same values we started with for the Bisection Method, 4 and 5, and we will let $x_0 = 4$ and $x_1 = 5$. Plugging those into the sequence to find x_2, and remembering that x_{n-1} means the previous value (x_1 in this case) and x_{n-2} means the value two steps ago (x_0 in this case), we get the following:

$$x_2 = 5 - f(4) * \frac{5 - (4)}{f(5) - f(4)}$$

Calculating $f(5)$ and $f(4)$, which are -48 and 32 respectively, our formula now becomes:

$$x_2 = 5 - 4 * \frac{5 - (4)}{-48 - (32)} = 4.4$$

Now, we find x_3, using the fact that we now know that $x_2 = 4.4$ and $x_1 = 5$. Plugging these values into the recursive formula to find x_3 gives us:

$$x_3 = 4.4 - f(4.4) * \frac{4.4 - 5}{f(4.4) - f(5)}$$

Which, if we evaluate $f(4.4)$ (which is 3.84), and recalling that $f(5)$ is -32, we get that:

$$x_3 = 4.4 - 3.84 * \frac{4.4 - 5}{3.84 - 32} \approx 4.4 + 0.064286 \approx 4.464286$$

Repeat, and find x_4, using the fact that we now know that $x_3 = 4.464286$ and $x_2 = 4.4$. Plugging these values into the recursive formula to find x_4 gives us:

$$x_4 = 4.464286 - f(4.464286) * \frac{4.464286 - (4.4)}{f(4.464286) - f(4.4)} \approx 4.449339$$

©Brian Gillispie, 2016.

Now we will find x_5, using the fact that we now know that $x_4 = 4.449339$ and $x_3 = 4.454286$. Plugging these values into the recursive formula to find x_5 gives us:

$$x_5 = 4.449339 - f(4.449339) * \frac{4.449339 - (4.454286)}{f(4.449339) - f(4.454286)} \approx 4.4494893$$

We will stop here. So, based on our work so far, we would say that the zero of our equation is about 4.4494893. In fact, if we plug into our equation 4.4494893, we get an output of $3.47 * 10^{-5}$, which means we are, after only five iterations, very close to our zero already!

Traditionally, the Secant Method is faster than the Bisection Method [13], however, unlike the Bisection Method, the Secant Method is not guaranteed to ever converge to the zero, even if the function is continuous [13]! Instead, the Secant Method can end up alternating between two values, neither of which are zero.

Also, while the Secant Method is usually faster than the Bisection Method, it is very possible to run the Secant Method five or six steps and still be very far from the zero. As a result, many modelers use the Bisection Method for a couple of steps to make sure they are close to the zero, then they switch over to the Secant Method at that point in time. Which method you use though is ultimately up to you to decide.

©Brian Gillispie, 2016.

9.1 Exercises

Find at least one of the zeros of the following functions. For these problems, you are free to use any valid method, even ones from previous Chapters.

1. $y = 25x - 200$

2. $y = 0.6x - 6000$

3. $y = -16x^2 + 100x + 4$

4. $y = -2.66x^2 + 67x + 10$

5. $y = -40x^2 + 24x + 2$

6. $y = x^3 - 3$

7. $y = x^3 - 3x^2 - 2x + 1.2$

8. $y = x^4 - 2x^2 - 16$

9. $y = 12x^5 - 60$

10. $y = 2x^4 - 3x^3 + 21x - 14$

11. $y = \ln(x + 10)$

12. $y = \log(45x + 100)$

©Brian Gillispie, 2016.

9.2: Domain Restricting

Back in Chapter 2, we discussed how all the models we are working with have the domain all real numbers. This means that for every model we have created so far, nothing is stopping someone from plugging in really large or really small inputs and getting a ridiculous prediction from the model. For this reason, the modeler often wishes to restrict the domain of their model, so that the model is not used to provide nonsense answers.

There are no hard and fast rules for how to restrict a domain. However, the best way to restrict the domain of a model is to look at the outputs, and ask ourselves "What is the largest and smallest output that should be allowed by the model?[130]". Then, we solve the equation to find the inputs that return these largest and smallest allowed values, and restrict the domain to only allow inputs between these two solutions. Let's demonstrate this with a few Examples.

Example 9.2.1: A particle is being heated in a laboratory. Initially the particle is 250 degrees, and it is warming up by 24 degrees per minute. At 586 degrees, the particle will turn to vapor and the current model will become invalid. Create a domain restricted model for this situation.

Solution to 9.2.1: To begin, we need to first create the model. We are given an initial value and a rate of change, which means we can create a linear model for this situation by the methods of Section 5.2. Using those methods, our initial linear model is, with x in minutes and y in temperature:

$$y = 24x + 250$$

Now, we need to restrict the model's domain. We know the model is invalid once the particle is 586 degrees, so we plug in 586 for y, since y represents the temperature of the particle. Then, we solve the following for x:

$$586 = 24x + 250$$

---

[130] If you are unsure what the largest and smallest values allowed should be, solving for when the model returns 0 is a good starting point, as very few models should ever return a negative number.

©Brian Gillispie, 2016.

Which, once we solve this for x, we get:

$$x = 14$$

Therefore, we conclude that we need to restrict the domain so that x is never larger than 14. We were given no other restrictions on the model[131], so this is the only one we will apply. Therefore, we conclude that our final model is:

$$y = 24x + 586, \quad x \leq 14$$

Example 9.2.2: A flock of birds has 150 birds in the year 2016, and the number of birds is increasing by 10% per year. It is estimated that the current rate of increase will hold until the year 2019, at which point it will drop a little. Create a domain restricted model for this situation.

Solution to 9.2.2: In this situation, we are given the number of birds, and the rate of increase, which means we need to use an exponential model. Using the rules of Chapter 6, and letting x be the years after 2016 and y being the number of birds, we end up with the following initial model:

$$y = 150 * 1.1^x$$

Next, we need to apply the restriction. We are told that the rate of increase will change in 2019, which will make the current model invalid. Therefore, we need to restrict x so that an input larger than the year 2019 is not plugged into the model. Since we defined x as years after 2016, this means that 2019 is represented by an input of 3 for x. Therefore, this means that we need x to be less than or equal to 3, which means our final domain restricted model is:

---

[131] Some say though that since the initial given data starts at $x = 0$, we should make it so that x is never less than 0 as well. We will not follow that rule here, as sometimes there is a valid reason for plugging in values of x less than 0.

©Brian Gillispie, 2016.

$$y = 150 * 1.1^x, \ x \leq 3$$

In the last two examples, we were given information that allowed us to find the domain restriction. Oftentimes that is not the case, and we have to use logic to find them. Let's demonstrate on another example a situation where we have to deduce the restrictions, because they are not stated.

Example 9.2.3: A ball is tossed into the air. The initial velocity of the ball is 50 ft/sec, and the initial height of the ball is 2.4 feet. If we assume $g = 32$ ft/sec^2, create a domain restricted model for this situation.

Solution to 9.2.3: First, we need to create the model. We are given an initial velocity, an initial height and a value for the acceleration of gravity, which means we need to use the height model from Chapter 7. Plugging in the given values as appropriate into the height equation gives us the following model:

$$y = -16t^2 + 50t + 2.4$$

Next, we need to restrict the domain of the model, but we were not given any restrictions to apply! This time we will have to think about it and see if there are any logical restrictions we should be applying. Turns out there are two. First, the model is invalid for inputs less than 0, because an input of less than 0 is before the ball is launched, which means the ball was doing something else (probably sitting at rest), making the model invalid for an input below 0. Secondly, the model is also invalid for any inputs where the resulting height is negative, as a negative height means the ball is passing through the ground. Therefore, we need to find when does our model return 0, and restrict the model to only allow inputs below that input which causes the model to return 0. In other words, we need to solve the following equation:

$$0 = -16t^2 + 50t + 2.4$$

©Brian Gillispie, 2016.

Using the quadratic formula from Chapter 7, we find that the positive input where this occurs is when $t = 3.17$. Therefore, we need to restrict our domain such that inputs are between 0 and 3.17, which makes our final model:

$$y = -16t^2 + 50t + 2.4, \quad 0 \leq t \leq 3.17$$

Notice in this scenario we ended up restricting the input such that a negative output was not possible. Usually this is done by finding the zeros of the model, then restricting the model appropriately, depending on what is happening either before or after the zero. In Example 9.3.3, we knew that inputs bigger than our zero would return a negative height, so we removed those inputs from our domain.

However, what do you do if you cannot find the zero because the model doesn't actually return zero, like for an exponential model? In those situations, it is best to pick some number that is really tiny and cut the model off there, in the absence of other information. We'll demonstrate this in our next example.

Example 9.2.4: A pollutant has entered a lake! As of July 2nd there are 5 liters of pollutant present, and the amount of pollutant is decreasing by 25% per day. Create a domain restricted model for this situation.

Solution to 9.2.4: First, create the model. Since we know the initial amount of the pollutant, and the rate of decrease, we will use the rules of Chapter 6 to create a model, where x is in days after July 2nd, and y is in liters of pollutant left. This yields the following model:

$$y = 5 * 0.75^x$$

Next, we need to apply the domain restrictions, but this time we were given none to apply in the problem. Nevertheless, we can deduce one restriction, as it makes no sense for the amount of the pollutant to ever be negative. But, we cannot solve for when this model returns 0, because exponential models never return 0, unless they have been shifted in some way. However, exponential models on a calculator may say the output is zero, due to the output being so small it

©Brian Gillispie, 2016.

is essentially zero. Since this usually happens near the point underflow occurs, we will now solve for what input values cause underflow to occur. Since underflow is around[132] $2.2 * 10^{-16}$, we will solve for when does the exponential model return $2.2 * 10^{-16}$. This means we need to solve the following equation:

$$2.2 * 10^{-16} = 5 * 0.75^x$$

Using the rules in Chapter 6 to solve this equation, we find that:

$$x = \frac{\ln(4.4 * 10^{-17})}{\ln(0.75)} \approx 130.91$$

Therefore, our domain restricted model, restricted in an attempt[133] to avoid outputs from the model returning zero due to underflow, is:

$$y = 5 * 0.75^x, \ x \leq 130.91$$

Unfortunately, the method in Example 9.2.4 does not always work, due to the value for underflow being different on different machines. For this reason, usually modelers decide in advance to say that outputs of a certain value for exponential models are essentially zero, and cut the model off there. Still, just be prepared for the possibility that any model that is restricted in an attempt to prevent answers from underflowing to zero occurring may still return zero, due to the type of software used.

---

[132] The actual number varies by calculator. However this is a good estimate [13] for most scenarios until such a point in time you are able to find the actual value that underflow occurs at.

[133] We say attempt, because on some calculators values close to 130 for an input may still return 0. There's no way to avoid that short of either cutting off at a much bigger output value (like 0.5), or customizing the model for each and every machine it is used on.

©Brian Gillispie, 2016.

9.2 Exercises

Create the appropriate domain restricted model.

1. A car is accelerating at 5 mph per second, and is currently going 15 mph. The car will stop accelerating at 65 mph. Create a linear domain restricted model for this situation.

2. In 2017 there were 240 rabbits, and the number of rabbits is increasing by 75% per year. If the rate of increase is only valid until 2021, create an exponential domain restricted model for this situation.

3. A tumor currently weighs 0.5 grams, and is decreasing by 45% per week due to treatment. Once the tumor weighs 0.1 grams it will be removed by surgery, making the model invalid. Create an exponential domain restricted model for this situation.

4. On July 1st there are 6000 customers in the waterpark, and the number of customers is expected to increase by 400 per day due to the upcoming holiday. However, the model is expected to only be valid for July 1st through July 5th, then everyone returns to work. Create a linear domain restricted model for the situation.

5. A pollutant has escaped into a river. Given there is 10 liters of pollutant in the river, and the amount of pollutant is decreasing by 90% per day, create an exponential domain restricted model if the model assumes outputs smaller than 0.01 liters is invalid.

6. Repeat Exercise 6, but this time assume outputs of $2.2 * 10^{-16}$ are invalid.

7. A ball is tossed into the air. If the ball was tossed with the velocity of 36 feet/sec, and an initial height of 10 feet, if we assume that $g = 32$ ft/sec^2, create an appropriate quadratic domain restricted model.

©Brian Gillispie, 2016.

8. At 8 am, it begins to rain, and at 9 am 1.4 inches of rain had fallen. Create a linear domain restricted model for this situation, if the storm is supposed to stop at 11 am.

9. An amusement park has 450 customers when it opens at 8 am, and 500 customers at 9 am. The park can hold 1000 customers. Create a linear domain restricted model for this situation.

10. Repeat Exercise 9, but this time, create an exponential domain restricted model.

11. A business has noticed that if they charge $30 for their newest software package they will make $180,000, if they charge $40 for their software package, they will make $210,000, and if they charge $50 for their software package they will make $225,000. Create a quadratic domain restricted model for this situation, if we assume that they have to make $150,000 from sales, else it is not worth their time to make and sell the product.

12. A wildlife reserve has 300 birds in the month of April, and 325 birds in the month of May. The reserve also knows that the birds migrate to the reserve in March, and they leave the reserve to go south in October. Create the linear restricted model for this situation.

13. For the data in Exercise 12, create a quadratic domain restricted model, if we also know that there are 340 birds present in the month of June.

14. For the models you found in Exercises 12 and 13, predict how many birds will be present in September, before the birds leave to fly south for the winter.

©Brian Gillispie, 2016.

9.3: Piecewise Models

Sometimes in the course of modeling it is necessary to use two different models to model the situation at various points in time. An example of this is a model of a bouncing ball that is thrown into the air. When the ball is initially tossed into the air, the height equation model will model its height until it hits the ground. However, after the ball hits the ground and bounces, another model is needed to model the height of the ball after the bounce, as the original model will claim that the ball is now under the ground. Such a scenario is modeled by what we call a piecewise model, or a model that splits into two or more different cases, and which case you apply depends on where in time you are currently modeling.

In this Section, we will discuss the idea of creating these piecewise models. We will begin first by defining what a piecewise model or equation is, as well as how to read them, then, we will discuss how to create them for our models.

9.3.1: Reading Piecewise Models

First, we will begin by defining a piecewise model, or a piecewise function, which we define below:

> *Definition 9.3.1: A **piecewise function** is a function defined by various subfunctions, with each subfunction applying on various intervals of the domain.*

Or, in other words, a piecewise function is broken into various pieces, and each piece applies for only part of the domain. To properly use a piecewise function, you will have to look at the notation carefully, and be sure to use the proper subfunction depending on where in the domain you are. The appropriate domain for the subfunction is always noted to the right of the subfunction itself, and will give the range of x that the subfunction applies to.

Let's now demonstrate this definition by evaluating a few piecewise functions.

Example 9.3.1: Given the following piecewise function:

$$f(x) = \begin{cases} 4x + 50, & 0 \leq x < 10 \\ 90, & x \leq 10 \end{cases}$$

Evaluate the function at $x = 4$.

Solution to 9.3.1: Notice that in this Example, our given function is made up of two subfunctions, the function $4x + 50$ and the function 90. Based on the rules noted to the right of each subfunction, the function $4x + 50$ applies only when x (or the input to the function) is between 0 and 10, with 0 included as well, and the subfunction 90 applies only when x (or the input to the function) is 10 or more.

Since our input is 4, which is between 0 and 10, this means we use the subfunction $4x + 50$ to evaluate our input. Plugging in our input, we get:

$$4(4) + 50 = 66$$

Therefore, when x is 4, the function returns 66.

Example 9.3.2: Evaluate the function in Example 9.3.1 for when $x = -1$.

Solution to 9.3.2: First, check and see which subfunction applies for our given input. Our input of -1 is not such that it is between 0 and 10, and our input is also not greater than 10. Therefore, none of the subfunctions apply to this input. When that happens, the input is undefined.

Be careful when using piecewise functions, as you will have to look closely sometimes to see if equal to is included in your subfunction. Let's demonstrate.

©Brian Gillispie, 2016.

Example 9.3.3: Given the following piecewise function:

$$f(x) = \begin{cases} -16x^2 + 64x, & 0 \leq x < 4 \\ 25, & x \geq 4 \end{cases}$$

Evaluate the function at $x = 4$.

Solution to 9.3.3: Be very careful here, as 4 is considered to be on the boundary of the domain of the subfunctions, so we need to see which subfunction includes inputs exactly equal to 4. By definition, the domain on the first subfunction is $0 \leq x < 4$, which does not include inputs of 4. Therefore, we cannot use the first subfunction to evaluate this input.

Looking at the second subfunction, we see that it is defined for inputs greater than or equal to 4, which means our input of 4 would use this subfunction. Therefore, we plug 4 into that subfunction and get 25.

Therefore, we conclude that when $x = 4$, this function returns 25.

9.3.2: Creating Piecewise Models

Now that we have seen how to read an existing piecewise model, it is time to look at how to create a piecewise model. To create a piecewise model, you need to create each individual model based on the conditions given, and specify the domain by each piece. Let's demonstrate.

Example 9.3.4: A modeler is observing a flock of ducks on a pond. If there are currently 35 ducks on the pond, and the number of ducks is increasing by 10% per month, but the pond can only hold 60 ducks, so once there are 60 ducks there will always be 60 ducks. Create a piecewise model for this situation.

Solution to 9.3.4: To begin, we need to create the model for the initial situation. We know there are 35 ducks, and the number of ducks is increasing by 10% per month, so by the rules of exponential models, our model is the following for now:

©Brian Gillispie, 2016.

$$y = 35 * 1.1^x$$

Now, we need to figure how to break this into pieces. We know that once the pond has 60 ducks, it will always have 60 ducks, no matter what. Therefore, we need to figure out what input causes the current model to return 60 (the maximum number of ducks allowed), and restrict that piece's domain based on that. To do this, we have to solve:

$$60 = 35 * 1.1^x$$

Solving this via the rules from Chapter 6, we find that:

$$x = \frac{\ln\left(\frac{60}{35}\right)}{\ln(1.1)} \approx 5.66$$

Therefore, the first piece of the model is:

$$y = 35 * 1.1^x, \quad 0 \leq x < 5.66$$

Next, we need to figure out the next piece. So far we know that once about 5.66 months pass (per our previous work), the pond has 60 ducks. Then, we also know that based on the given information, at this point in time, the pond will always have 60 ducks. This means that for any input higher than 5.66, the model has to return 60, or, our second piece of the model is:

$$y = 60, \quad 5.66 \leq x$$

Putting this together, this means our piecewise model for this situation is:

$$y = \begin{cases} 35 * 1.1^x, & 0 \leq x < 5.66 \\ 60, & 5.66 \leq x \end{cases}$$

Sometimes, when creating the piecewise model, we will have to shift our inputs for the second (or later) subfunctions so that the proper output is returned. Let's demonstrate.

Example 9.3.5: A ball is tossed into the air on Earth ($g = 32$ ft/sec^2) with an initial velocity of 64 ft/sec, and an initial height of zero feet. Once the ball hits the ground, it will bounce into the air again, but with a velocity of 16 ft/sec. When it hits the ground the second time, it stays on the ground. Create a piecewise model for this situation.

Solution to 9.3.5: Start out by creating the model for the initial situation. We know that initially, the ball is thrown into the air with a given velocity and given height, which means we need to use the height equation to model this. Using our height equation from Chapter 7, our model for the first subfunction is:

$$y = -16x^2 + 64x$$

Next, we need to find when this model hits the ground so that we know when to switch to the second subfunction. Using the quadratic formula from Chapter 7, we find that this ball hits the ground when $x = 4$. Therefore, our first subfunction will apply in the interval $0 \leq x < 4$

Now, create the subfunction for when the ball bounces for the first time. We know that the velocity changes to 16ft/sec once it bounces, and, even though we are not given a new initial height, since the ball hit the ground to bounce, it means it will have an initial height of zero here. This means that our second subfunction is:

$$-16x^2 + 16x$$

However, if we try to use that subfunction to model the situation after the bounce, it will say that the ball is already in the ground! This is because the height equation expects the model to start at time zero, and this time we started at time four. To fix this, we need to replace x with $(x - 4)$ in our model to shift the model to account for our new input. This means our correct subfunction for this situation is:

©Brian Gillispie, 2016.

$$y = -16(x-4)^2 + 16(x-4)$$

And finally, we need to find when that submodel hits the ground (again), as that is when the ball stays on the ground. If we solve for that point using the quadratic formula on our subfunction, we find that this ball hits the ground when $x = 4$ and when $x = 5$. This tells us that the ball hits the ground the second time when $x = 5$, as the first solution to the quadratic equation is telling us when the ball bounced off the ground. Therefore, this means our second subfunction is the following:

$$y = -16(x-4)^2 + 16(x-4), \quad 4 < x \leq 5$$

We are still not done, as we need a third subfunction to represent when the ball hits the ground the second time and stays there. As staying on the ground is represented by a height of zero, this means that after the ball hits the ground at time 5, the model needs to always return 0. This is represented by our third subfunction being the following:

$$y = 0, \quad 5 < x$$

Therefore, our final model for this situation is the following:

$$y = \begin{cases} -16x^2 + 64x, & 0 \leq x \leq 4 \\ -16(x-4)^2 + 16(x-4), & 4 < x \leq 5 \\ 0, & 5 < x \end{cases}$$

To use these piecewise models to make predictions, work with them as we did when we evaluated them. However, do be aware that if you need to solve for when a piecewise function returns a certain y value, you will have to solve each subfunction for that y value, and return any answer that is still in the subfunction's domain. Let's demonstrate.

©Brian Gillispie, 2016.

Example 9.3.6: A new tech gadget has been released! Initially, 1000 copies of the gadget sell on the date of release (August 1st), and the number of copies selling is increasing by 100 per day. However, once 2000 copies sell, projections say that sales will take off rapidly at a linear pace, with sales peaking at 3500 copies 3 days after the device sold 2000 copies, but then sales will drop rapidly at an exponential rate, with sales dropping by 20% per day. On which days will the device sell 1500 copies?

Solution to 9.3.6: We have quite a few points to work with here, and all of them use different models. Start out by fist creating the model for the initial given info, which is based on the fact that on August 1st, 1000 copies are released, and the sales are increasing by 100 per day. Using that information, and what we learned in Chapter 5, we find that our first model is, with x days after August 1st, and y copies sold:

$$y = 100x + 1000$$

Next, we need to find when the subfunction that we just found ends. We know that things change when 2000 copies sell, so plug in 2000 for y and solve. If we do that, we find that we sell 2000 copies when x is 10, so that means the first subfunction is for the domain $0 \leq x < 10$

Next, create the subfunction for when the model increases from 2000 to 3500. We are told that it takes off at a linear pace, so we will use a linear function to connect those two points. However, we need to figure out the x values for these points. We know that an input of 10 should return 2000, which makes the first point $(10, 2000)$, and the sales will reach 3500 3 days later. Since x represents days after August 1st, this means that the second point is $(13, 3500)$. Therefore, if we create the line which connects the points $(10, 2000)$ and $(13, 3500)$, we find that the connecting line is:

$$y = 500x - 3000$$

©Brian Gillispie, 2016.

Also, since we know that this subfunction only applies between those two points[134], we know then that the domain of this subfunction is $10 \leq x < 13$.

Next, create the next subfunction for when the sales decrease. We know that from the point $(13, 3500)$ the sales drop by 20% per day at an exponential rate. Therefore, using the rules from Chapter 6, this means our subfunction for this situation is the following, with the subfunction being in effect if our inputs are 13 or more based on the given information:

$$y = 3500 * 0.80^{(x-13)}$$

Notice we have to use $x - 13$ and not x, as the model in Chapter 6 starts at an input of 0, so we have to fix it so that an input of 13 returns 3500.

With all of the subfunctions known, we can create our model. Our model for this situation is the following:

$$y = \begin{cases} 100x + 1000, & 0 \leq x < 10 \\ 500x - 3000, & 10 \leq x < 13 \\ 3500 * 0.80^{(x-13)}, & 13 \leq x \end{cases}$$

Finally, we can answer the question asked. We wish to know on which days the sales of the device will be 1500. To answer that we have to set all three pieces of the model equal to 1500, and see if those answers are in the domain. Doing this for the first subfunction, we find that:

$$1500 = 100x + 100$$

Occurs when:

$$x = 5$$

---

[134] This is one advantage to using the two endpoints to create the model, as we already know at what inputs the model has to change to the next subfunction.

©Brian Gillispie, 2016.

This is in the domain for the first subfunction, as the subfunction is defined on the interval $0 \leq x < 10$, and since 5 is in that interval, 5 is a valid answer to the problem.

Next, solve the second subfunction. We need to find when:

$$1500 = 500x - 3000$$

This occurs when:

$$x = 9$$

However, 9 is outside the interval defined for the subinterval of $10 \leq x < 13$, so this is not a valid solution to the problem.

And finally, we solve the third subfunction, or, we solve:

$$1500 = 3500 * 0.8^{(x-13)}$$

Solving this equation for x, we find that:

$$x = 16.80$$

Since the subfunction is defined in the interval $13 \leq x$, and our answer is in this interval, this is also a valid solution to the problem. Therefore, we conclude that there will be 1500 copies sold when $x = 5$ and $x = 16.80$, or, since x is defined as days after August 1st, August 6th and about August 18th respectively.

©Brian Gillispie, 2016.

9.3 Exercises

For Exercises 1 - 9, evaluate the given piecewise functions at the given points.

1. Evaluate $f(x) = \begin{cases} 3x - 12, & 0 \leq x < 4 \\ x - 4, & 4 \leq x \end{cases}$, at $x = 2$

2. Evaluate the function in Exercise 1 at the point $x = 6$

3. Evaluate $f(x) = \begin{cases} 100 * 1.05^x, & 0 \leq x < 2 \\ 110.25 * 1.03^{(x-2)}, & 2 \leq x \end{cases}$, at $x = 7$

4. Evaluate $f(x) = \begin{cases} 0, & 0 \leq x \leq 10 \\ -16(x - 10)^2 + 132(x - 10), & 10 < x \end{cases}$, at $x = 3$

5. Evaluate the function in Exercise 4 at $x = 12$

6. Evaluate the function in Exercise 4 at $x = -1$

7. Evaluate $f(x) = \begin{cases} 100 * 1.2^x, & 0 \leq x \leq 5 \\ 249 * 0.99^{(x-5)}, & 5 \leq x < 9 \\ 239, & 9 < x \end{cases}$, at $x = 2$

8. Evaluate the function in Exercise 7 at $x = 6$

9. Evaluate the function in Exercise 7 at $x = 14$

©Brian Gillispie, 2016.

10. Given the function $f(x) = \begin{cases} 100 * 1.2^x, & 0 \leq x \leq 5 \\ 249 * 0.99^{(x-5)}, & 5 \leq x < 9 \\ 239, & x \leq 9 \end{cases}$, for what inputs does the model return 241?

11. A colony of bacteria is in a tube, and there is currently 400 cells in the tube, and the number of cells is increasing by 5% per minute. If once there is 800 cells in the tube, there will always be 800 cells in the tube, create a piecewise function for this situation.

12. A ball is launched in a cannon! If the ball is launched at an initial velocity of 300 ft/sec, and an initial height of 2 feet, if we assume that $g = 32$ ft/sec^2, create a piecewise model for this situation, if the ball, once it hits the ground, it stays on the ground.

13. Repeat Exercise 12 if the ball bounces once when it hits the ground, with a velocity of 60 ft/sec. Assume the ball only bounces once, then stays on the ground.

14. Repeat Exercise 12 if the ball bounces twice when it hits the ground, with the first bounce having a velocity of 150 ft/sec, and the second bound having a velocity of 55 ft/sec. Assume the ball stays on the ground after the second bounce.

15*. Repeat Exercise 12 if the ball bounces with 30% of the velocity it had before on each bounce. Assume if the ball ever has a velocity of 1 ft/sec or less that it stays on the ground from there on.

16. You have been hired for a new job! If your starting salary is $50000 a year, and at the end of the first year you will get a raise of $5000, then all future raises will increase your previous year's salary by 3%, calculate what your salary will be 5 years from now.

17. Using the data in Exercise 16, calculate when you will be making $60000 a year.

©Brian Gillispie, 2016.

18. A bunch of rabbits have escaped and made their way to an island! If we initially have 250 rabbits, and each year the number of rabbits are increasing by 200% until there are 30000 rabbits, then they will increase by 75% per year, create a piecewise model for this situation, and use it to predict how many rabbits will be present on the island 10 years from now.

19. Using your model in Exercise 18, predict how many years until there are 25000 rabbits on the island.

20. Once there are 50000 rabbits on the island in Exercise 18, wolves are introduced to the island, and the wolves are decreasing the rabbit population by 5% per year. Using both this information and your model in Exercise 18, predict how many rabbits will be present 30 years from the time the rabbits escaped.

21. Currently, your business is supplied with corn at the rate of $1 per pound, but due to rising costs, the cost of corn is increasing by 10% per year. Once the cost of corn reaches $2 per pound, all predictions say that the cost of corn will only increase by 2% per year from then on. Create a piecewise model and use it to predict the cost of corn 10 years from now.

22. A town has 10000 people in the year 2020, but due to the loss of a gold mine, people are leaving the town. If the population of the town is decreasing by 20% per year, and all estimates indicate that this will continue for 4 years until a new technology company opens a business there! Once that happens, the population of the town will increase by 7% per year. Create a piecewise function, and use it to predict how many people will be in the town in the year 2030.

23. Use your model in Exercise 22 to predict in which years the town will have 5000 people.

©Brian Gillispie, 2016.

9.4: Parametric Models

Most real world situations that we wish to model cannot be modeled with only a single model, but are instead modeled by two or more models that are parameterized by time. These models are called parametric models. In this Section, we will discuss how to work with these parametric models, and will then work with some commonly known parametric models as well.

9.4.1: Definition of Parametric Models

A parametric model is a model where two or more variables are parameterized by a relating variable, usually time. An example of a parametric model is:

$$x = 2t$$
$$y = -16t^2 + 64t + 5$$

In both equations, x and y depend on time (t) in the model, and the variable t is said to be the parameter in the model.

To evaluate models that have been parameterized, plug in the given value of t and use that to find what x and y (and the other variables, if more are present) are. Let's demonstrate:

Example 9.4.1: Given the following parametric model:

$$x = 4t$$
$$y = -16t^2 + 64t + 2$$

Evaluate the model for $t = 4$.

©Brian Gillispie, 2016.

Solution to 9.4.1: To evaluate the model we need to plug in our given value in for t in both equations. If we do that, we end up with:

$$x = 4(4) = 16$$
$$y = -16 * 4^2 + 64 * 4 + 2 = 2$$

Therefore, when t is 2, x is 16 and y is 2.

Example 9.4.2: Given the following parametric model:

$$x = 44 * t$$
$$y = -16t^2 + 42t + 3$$
$$z = -0.5t + 2$$

Evaluate the model for $t = 1.2$

Solution to 9.4.2: To evaluate the model when t is 1.2, we need to plug 1.2 into all three of the equations this time. If we do that, we end up with:

$$x = 44 * 1.2 = 52.8$$
$$y = -16(1.2)^2 + 42 * 1.2 + 3 = 30.36$$
$$z = -0.5 * 1.2 + 2 = 1.4$$

Therefore, when t is 1.2, x is 52.8, y is 30.36, and z is 1.4.

Sometimes, with parametric models, it will be necessary to solve for when one model is equal to a certain value, and then plug in that time into the other equations. Let's demonstrate:

©Brian Gillispie, 2016.

Example 9.4.3: Given the following parametric models to represent the height of a ball:

$$x = 22 * t$$
$$y = -16t^2 + 16.4t + 2$$

Find what x is when y is 0. Assume t has to be positive.

Solution to 9.4.3: To figure out what x is when y is 0, we need to first set the y equation equal to 0 and solve for t. Then, once we know that value of t, we plug it into our x equation to find x at that point in time.

First, let's find at what time y is equal to 0. Setting y equal to 0 gives us:

$$0 = -16t^2 + 16.4t + 2$$

This is a quadratic equation that we need to solve. Solving it via the rules in Chapter 7, we find that this occurs when:

$$t = -0.11, \quad 1.13$$

However, since we were told to assume t is positive, we will only use the answer 1.13. Plugging that into the x equation for y, we find that then when t is 1.13, x is:

$$x = 22.4 * 1.13 = 25.312$$

Therefore, when y is 0, x is 25.312.

©Brian Gillispie, 2016.

9.4.2: A Common Parametric Model: Modeling Height and Distance

Now we will turn our attention to one of the most common parametric models used, the parametric equations of height and distance of an object that is thrown into the air at an angle. If we assume there is no air resistance, it can be shown that the equations to model the motion of the object can be represented by the following:

*Definition 9.4.1: The **parametric equations** for the height and distance of a thrown object after t seconds can be given as follow, with x the forward distance of the object, and y the height of the object:*

$$y = -\frac{g}{2}t^2 + v_0 * \sin(\theta) * t + h_0$$

$$x = v_0 * \cos(\theta) * t$$

Where:

g = acceleration due to gravity

v_0 = initial velocity

h_0 = initial height

θ = angle the object is thrown at

The y equation should be familiar from Chapter 7, the only difference is now we are including the angle the object is thrown at. Now, let's work a few Examples where we set up problems with these parametric equations before proceeding.

Example 9.4.4: A rock is launched out of a cannon at an angle of 23 degrees, with an initial velocity of 350 ft/sec, and an initial height of 6 feet. If we assume that $g = 32$ ft/sec², create the parametric equations for this situation.

Solution to 9.4.4: With the data given, our initial velocity (v_0) is 350, our initial height (h_0) is 6, our g is 32, and our angle θ is 23. If we plug this all into the equations, we get:

©Brian Gillispie, 2016.

$$y = -\frac{32}{2}t^2 + 350 * \sin(23) * t + 6$$
$$x = 350 * \cos(23) * t$$

Which, can be simplified to[135]:

$$y = -16t^2 + 136.76 * t + 6$$
$$x = 322.18 * t$$

To work with height and distance problems involving parametric equations requires the modeler to usually have to solve for when one of the equations (height or distance) hits a specified value, then use that value of time in the other equation. Therefore, when working these problems, it is important to remember that x stands for forward distance, and y stands for height. Keeping those two straight will be critical as we work the remaining problems in this Section.

Next, let's work a few Examples to show how to use these equations to calculate certain heights and distances.

Example 9.4.5: A football is kicked with an angle of 18 degrees, an initial velocity of 76 ft/sec, and an initial height of 3 feet. If we assume that the acceleration due to gravity is 32 ft/sec^2, how far forward does the football travel before it hits the ground?

Solution to 9.4.5: First, we will want to set up our equations. We know that our ball travels at an initial velocity (v_0) of 76, our initial height (h_0) is 3, our g is 32, and our angle θ is 18. Plugging those into our equations gives us (after simplifying):

$$y = -16t^2 + 23.49 * t + 3$$
$$x = 72.28 * t$$

---

[135] Be sure your calculator is in degree mode before simplifying these! By default, most calculators are in radian mode when first bought, and radian mode will not work for these problems unless you convert your angles to radians first.

©Brian Gillispie, 2016.

Next, we need to figure out what we need to calculate. In this problem, we wish to know how far forward the ball travels before it hits the ground. Since travels means forward distance, or x, and ground involves height, or y, and hitting the ground means that the height is zero, this can be thought of as asking what is x when y is zero. Therefore, we set y equal to zero and solve for t. Doing that gives us:

$$0 = -16t^2 + 23.49 * t + 3$$

If we solve this equation, we find that:

$$t = -0.12, 1.59$$

However, since negative time makes no sense in the context of this problem, we discard the answer of -0.12, which means that we conclude that the ball hit the ground 1.59 seconds after it was thrown. Plugging that value for t into the x equation allows us to find the distance the object travels, which, if we do that, we get:

$$x = 72.28 * 1.59 = 114.93$$

Therefore, we conclude that the football traveled 114.93 feet forward before it hit the ground.

Example 9.4.6: This is it, the big moment! The bases are loaded, it is the bottom of the 9th inning, and the count is 3 balls and 2 strikes. The pitcher throws you a fastball, which you proceed to hit as hard as you can. Given that the ball was hit with a velocity of 143 feet/sec, an initial height of 3 feet, and an angle of 14 degrees, and an acceleration due to gravity of 32 ft/sec^2, does the baseball clear a home run fence that is 400 feet away from you and 12 feet tall?

©Brian Gillispie, 2016.

Solution to 9.4.6: First, we need to set up our parametric equations. We know that the baseball travels at an initial velocity (v_0) of 143, the initial height (h_0) is 3, our g is 32, and our angle θ is 14. Plugging those into our equations gives us (after simplifying):

$$y = -16t^2 + 34.59t + 3$$
$$x = 138.75t$$

Next, we need to figure out what we need to calculate. This time, the problem is asking us to figure out if once the ball travels 400 feet forward, is the baseball more than 12 feet in the air? To answer that, we need to know how long it takes the baseball to travel 400 feet. Since x represents forward distance, we will figure out how long it takes the ball to travel 400 feet, then see if the ball has the desired height at that point in time. Plugging in 400 for x into the x equation gives us:

$$400 = 138.75t$$

Which, if we solve for t, we get:

$$t = 2.88$$

That tells us that it takes the baseball 2.88 seconds to get to the homerun fence. Now, we need to see if the baseball has the height necessary to clear the fence. Since height is represented by y in these equations, we will plug in our value of t we found into the y equation and see how high the baseball is at this point in time. Doing that gives us:

$$y = -16(2.88)^2 + 34.59(2.88) + 3 \approx -30.09$$

This is less than our desired height of 12 (the height of the fence), so we conclude that the baseball does not clear the homerun fence. Instead, you hit a hard line drive.

©Brian Gillispie, 2016.

9.4 Exercises

For Exercise 1 - 4, evaluate the given parametric model

1. Evaluate the given model at $t = 1.4$

$$x = 10t$$
$$y = -16t^2 + 55t + 26$$

2. Evaluate the given model at $t = 20$

$$x = 3t - 2$$
$$y = 100 * 1.05^t$$

3. Evaluate the given model at $t = 4.8$

$$x = 99t$$
$$y = -16t^2 + 144t + 10$$

4. Evaluate the given model at $t = 0.7$

$$x = 5t$$
$$y = -16t^2 + 77t + 1$$
$$z = t$$

©Brian Gillispie, 2016.

For Exercise 5 onward, set up and solve the given parametric distance/height models.

5. A ball is thrown into the air with an initial velocity of 10 ft/sec, an initial height of 2, and an angle of 55 degrees. If we assume $g = 32$ ft/sec^2, how far forward does the ball travel before it hits the ground?

6. Repeat Exercise 5, but this time let $g = 12.2$ ft/sec^2

7. A baseball is hit with an initial velocity of 99 ft/sec, an initial height of 3.6 feet, and an angle of 78 degrees. If we assume that $g = 32$ ft/sec^2, how far forward does the baseball travel before it hits the ground?

8. Repeat Exercise 7, but this time the angle the baseball is hit at is 4 degrees.

9. Repeat Exercise 7, but this time the angle the baseball is hit at is 100 degrees. Why do you think you got the answer you did for this problem?

10. A dart is thrown at a dartboard as hard as possible. Given that the dart is thrown at 100 ft/sec, an initial height of 4.7 feet, and an initial angle of 2 degrees, if we assume that $g = 32$ ft/sec^2, how high is the dart when it hits the dartboard that is 5 feet in front of you?

11. You hit a golf ball on the moon! Given that you hit the golf ball with an initial velocity of 125 ft/sec, an initial height of 0 feet, an initial angle of 36.5 degrees, if we assume that $g = 5.3$ ft/sec^2, how far does the golf ball travel?

12. In Exercise 11, if there is a 160 foot tall cliff that is 500 feet in front of where you hit the golf ball on the moon, does the golf ball clear the cliff? Why or why not?

©Brian Gillispie, 2016.

13. A baseball is hit with an initial velocity of 112 ft/sec, an initial height of 4.2 feet, and an angle of 26 degrees, and the acceleration due to gravity is 32 ft/sec². If the home run fence is 420 feet away, and 10 feet high, does the ball clear the fence?

14. A foul ball is hit with an initial velocity of 98 feet/sec, an initial height of 4.4 feet, and an angle of 88 degrees. If the upper deck is 40 feet high and 50 feet in front of where the ball is going, does the ball make it into the upper deck to become a souvenir for a happy fan?

15. A field goal kicker is attempting to make a game winning field goal. If the ball is kicked with an initial velocity of 76 ft/sec, an angle of 36 degrees, and no initial height, and if the acceleration due to gravity is 32 ft/sec², does the kick clear the crossbar which is 120 feet in front of the kicker and 10 feet high?

16. In the same scenario as in Exercise 15, a defender is 9 feet away when the ball is kicked, and reaches as high as they can in an attempt to block the kick. If their hand is 7.2 feet in the air and in the path of the football, and if we assume that if the ball does not clear the hand the kick is blocked, is this kick blocked?

17. A field goal kicker is about to attempt a winning field goal. If the kick has an initial velocity of 82 ft/sec, an angle of 37 degrees, and no initial height, if we assume that the acceleration due to gravity is 32 ft/sec², and the crossbar is 169 feet away and 10 feet high, does the kick clear the crossbar?

18. In the same scenario as in Exercise 17, a breeze is blowing across the field, and introduces skew. The skew can be modeled by the equation $z = -4.5t + 3.4$. If we assume that any z value in the interval of $-9.25 \leq z \leq 9.25$ means the kick is inside the crossbar, is the kick in Exercise 17 inside the crossbar or not?

©Brian Gillispie, 2016.

Chapter 9 Summary

9.1: In this Section, we discussed various methods of finding zeros of equations. In particular, we discussed the bisection method, and the secant method.

The Bisection Method is based on the idea that if we know a function is negative for one input and positive for another, we can take the midpoint of that interval, and then, depending on the value of the midpoint, cut the interval in half and repeat.

The Secant method is based on a recursive method, which is:

$$x_n = x_{n-1} - f(x_{n-1}) * \frac{x_{n-1} - x_{n-2}}{f(x_{n-1}) - f(x_{n-2})}$$

9.2: In this Section, we discussed the idea of domain restricting our models. Usually models are domain restricted based on what inputs the model is valid for, though sometimes judgment is needed on these values.

9.3: In this Section, we discussed the idea of creating a piecewise model, for when a new model is needed for the situation being modeled. In order to do this, we need to be able to figure out at which point the model changes, and set up our model accordingly.

©Brian Gillispie, 2016.

9.4: In this Section, we discussed parametric models, or models where the x and y variables (and sometimes more) are all based on a parameter t. Then, we discussed how we can use a common parametric model to model the distance and height of an object. In general, the height (y) and distance (x) of an object can be parameterized by the following equations:

$$y = -\frac{g}{2}t^2 + v_0 * \sin(\theta) * t + h_0$$

$$x = v_0 * \cos(\theta) * t$$

Where:

$$g = acceleration\ due\ to\ gravity$$
$$v_0 = initial\ velocity$$
$$h_0 = initial\ height$$
$$\theta = angle\ the\ object\ is\ thrown\ at$$

Appendix A: Challenge Problems

Enclosed here are some hard modeling problems if you wish to challenge yourself. To work these problems successfully, you will need to use all of the material from all of the Chapters. Be sure to explain in full detail why you use the method that you do, as oftentimes there will be two or more ways to successfully work the problem.

©Brian Gillispie, 2016.

©Brian Gillispie, 2016.

Challenge Problem #1:

A snowplow has a top speed of 50 mph, but for every inch of snow on the road, the top speed of the snowplow decreases by 4.5 mph. You need to plow a highway that is 450 miles long, and there is 6.2 inches of snow on the highway. How long will it take you to plow this highway?

Challenge Problem #2:

You have decided to make a video game! In order to make this game, you will need to hire an artist for $70000 per year, another artist for $55000 per year, a musician for $70000 per year, another musician for $55000 per year, and a marketer for $65000 per year, but the marketer will only be on staff for the last 6 months of production.

Currently you estimate that the game will take you 3 years to make, but you can hire on an assistant which will reduce the time to make by 1 year, but will cost you an extra $200000 total for the two years you would need the assistant for. Should you hire on the assistant? Why or why not?

Challenge Problem #3:

A ball is thrown as hard as possible at an initial velocity of 110 ft/sec, an initial height of 4.6 feet, and an angle of 44 degrees, and the acceleration due to gravity is 32 ft/sec^2. If, once the ball hits the ground, it will bounce once, with an initial velocity of 44.3 ft/sec, then, once it hits the ground the second time, it will roll an additional 26 feet before coming to a stop, what is the entire distance that the ball travels forward before it comes to a stop in this situation?

©Brian Gillispie, 2016.

Challenge Problem #4:

In an attempt to model and predict when a business needs to expand, they record the following data on the number of customers they get per month. In January 2015, they had 7000 customers. In February 2015, they had 6885 customers. In March 2015, they had 7200 customers. In April 2015, they had 7400 customers. In May 2015, they had 7350 customers. In June 2015, they had 7960 customers.

The business will need to expand once the business hits 10500 customers regularly. Predict when that will happen.

Challenge Problem #5:

Research and devise a way to properly model skew with z in a parametric model. Use what is known about x and y in Section 9.4 to get you started, and let z represent how far the object moves side to side. Things to consider in your model are how the initial velocity of the object applies to the skew, as well as how much spin is put on the object will affect the skew.

Challenge Problem #6:

Record the high temperatures for your city for the last twelve days. Use those data points to make a reasonable prediction for what the high temperature will be five days later. How does your temperature compare with the official prediction given by your local weather?

Extra: Calculate the error of your prediction once the real high temperature is known. What factors affected the high temperature that your model forgot to account for?

©Brian Gillispie, 2016.

Challenge Problem #7:

A college currently has 1700 students for the Fall 2015 term, and projects to have 1778 students for the Fall 2016 term. Then, for the Fall 2018 term, they project to have 1902 students, and for the Fall 2019 term, they project to have 1925 students. The board of regents needs to build a new dorm, and they need it before the school has 2050 students present. You have three proposals on hand for building the dorm, from three different companies:

Company A: Says they can build the dorm for 12 million for Fall 2020, but for each extra year you give them the dorm cost will go down by 2.7 million dollars, to a minimum of 7 million dollars.

Company B: Says they can build the dorm for 12.5 million for Fall 2020, but for each extra year you give them the dorm cost will go down by 2.9 million dollars, to a minimum of 6.5 million dollars.

Company C: Says they can build the dorm for 13 million for Fall 2020, but for each extra year you give them the dorm cost will go down by 3.1 million dollars, to a minimum of 6 million dollars.

With all of this info, the board of regents needs you to propose what they should do. Which proposal should the company accept, with the understanding that it is mandatory the dorm is built before the college has 2050 students, and with the understanding that your goal is to build the dorm as cheaply as possible?

©Brian Gillispie, 2016.

©Brian Gillispie, 2016.

Appendix B: Answers to Selected Exercises

For some problems, there is more than one correct answer. In those cases we only included one of the correct answers. Also be advised that our answers are calculated with no to minimal rounding until the last step. If you round on the way, your answers will differ slightly.

1.1 Exercises

As these answers depend heavily on the opinion of the reader, no answers can be provided.

1.2 Exercises

1. 40

3. 48

5. 3932180

7. 2

9. 1244.12

11. 50.1

13. 59

15. 45

17. 1

19. 57.88125

21. 264

23. 2.7

25. 0

27. Too big to calculate

1.3 Exercises

1. 150 minutes

3. 2 hours

5. 600 seconds

7. 204 hours

9. 132 inches

11. 30 feet^2

13. 8 mlies^2

15. 32.5 $\frac{\text{cars}}{\text{stoplight}}$

17. 5 hours

19. 15 pizzas

21. $44.95

23. $8400

1.4 Exercises

1. Error: 0.1, Relative Error: 1.27%

3. Error: −1, Relative Error: −6.25%

5. Error: 2, Relative Error: 0.57%

15.

X	Y	Calculated Value	Absolute Error
0	0	0	0
1	34	32	2
2	28	32	4
3	0	0	0
Sum			6

17.

X	Y	Calculated Value	Absolute Error
1	10	10	0
2	4.5	5	0.5
4	2	2.5	0.5
Sum			1

19.

X	Y	Calculated Value	Absolute Error
0	100	100	0
5	130	130	0
10	168	169	1
15	206	219.7	13.7
20	233	285.61	52.61
Sum			67.31

21. $y = -16x^2 + 48x + 100$

23. $y = 9x + 103$

1.5 Exercises

M1. 396 inches^2

M3. 615.75 feet^2

M5. 20 feet^2

M7. 28.27 miles^2

M9. 75 feet^2

©Brian Gillispie, 2016.

M11. $135

M13. $30

D1: 40 hours

D3: 10.42 seconds

D5: 153.85 hours

D7: About 0.42 hours, or 25.2 minutes

D9: 8.33 meters per second

D11: 2200 feet

D13: 7.2 miles

2.1 Exercises

1. Quantitative

3. Categorical

5. Quantitative

7. Categorical

9. Answers vary

11. 2562

13. 12.67

2.2 Exercises

1. x is year, y is number of ducks. $(2000, 44)$, $(2010, 56)$

3. x is time in am, y is number of pigeons. $(10, 67)$, $(11, 69)$

5. x is time in pm, y is cells of bacteria. $(5, 1000)$, $(6, 2000)$

7. x is time in years, y is number of wolves. $(0, 25)$, $(4, 30)$

9. x is time in seconds, y is height in feet. $(0, 0)$, $(1, 72)$

11. x is time in years, y is number of salmon. $(0, 10000)$, $(1, 11000)$

13. x is time in days, y is height of river in feet. $(0, 12)$, $(1, 12.5)$

15. x is time in seconds, y is height in feet. $(0, 0)$, $(1, 10)$, $(2, 2)$

17. x is year, y is number of rabbits. $(2020, 32)$, $(2021, 675)$, $(2022, 4780)$, $(2023, 23417)$

2.3 Exercises

1. 55.6

3. 41.65

5. 69.80 geese

7. 5.42 inches

9. 0.57 calories

11. $(0, 325)$, $(4, 305)$, $(8, 287)$, $(12, 274)$

13. $(17, 5.4)$, $(18, 5.2)$, $(19, 5.5)$, $(20, 6.1)$, $(21, 6.0)$

15. $(10, 1000)$, $(12, 1075)$, $(14, 1150)$, $(16, 1269)$, $(19, 1501)$

17. $(300, 1)$, $(360, 1.075)$, $(420, 1.150)$, $(480, 1.269)$, $(570, 1.501)$

19. $(106, 122)$, $(107, 178)$, $(108, 115)$, $(109, 99)$, $(111, 350)$

21. $(2.67, 12.9)$, $(3, 11.5)$, $(3.33, 11.6)$, $(3.67, 10.2)$, $(4, 9.6)$, $(4.33, 9.9)$

23. $(0, 12.9)$, $(0.5, 11.5)$, $(1, 11.6)$, $(1.5, 10.2)$, $(2, 9.6)$, $(2.5, 9.9)$

©Brian Gillispie, 2016.

2.4 Exercises

1. $(1, 72), (2, 55), (3, 60), (4, 102), (5, 98)$, with Monday $= 1$, Tuesday $= 2$, etc.

3. $(12, 150000), (13, 167900), (14, 177090), (15, 172950)$, with $12 =$ Jan 12th, $13 =$ Jan 13th, etc.

5. $(21, \text{Vanilla}), (22, \text{Chocolate}), (23, \text{Strawberry})$, with $21 =$ July 21st, $22 =$ July 22nd, etc.

7. $(0, \$50), (1, \$55), (2, \$48)$, with Cinderella $= 0$, Beauty and the Beast $= 1$, RENT $= 2$.

9. $(0, 3), (1, 7), (2, 1), (3, 0)$, with $0 =$ Dec. 2014, $1 =$ Jan. 2015, etc.

11. $(29, 325), (30, 299), (31, 311), (32, 349)$, with $29 =$ June 29th, $30 =$ June 30th, $31 =$ July 1st, $32 =$ July 2nd

13. $(0, 14), (1, 7), (2, 0), (3, 0), (4, 2)$, with $0 =$ Oct. 2016, $1 =$ Nov. 2016, $2 =$ Dec. 2016, $3 =$ Jan. 2017, $4 =$ Feb 2017

15. $(0, 56), (3, 90), (6, 69), (9, 24), (12, 48)$ with $0 =$ Mar. 2015, $3 =$ June 2015, $6 =$ Sept. 2015, $9 =$ Dec. 2015, $12 =$ Mar. 2016

2.5 Exercises

1. $(-\infty, \infty)$

3. All points but $-\frac{7}{2}$

5. $[10, \infty)$

7. 18

9. 13.84

11. DNE

13. 90.8

15. 1964.23

©Brian Gillispie, 2016.

17. DNE

19. DNE

21. 24.79

25. $(-\infty, \infty)$

27. $(325, \infty)$

29. $[9, 75]$

2.6 Exercises

As these are research essays, no answers can be provided.

3.1 Exercises

1. TC -30, AVC -6

3. TC 50 frogs, AVC 5 frogs/month

5. TC 960 crickets, AVC 240 crickets/day

7. TC $300, AVC $10.71/year

9. TC $-$$0.20, AVC $-$$0.0286/day

11. AC 1.6, GC 1.22

13. AC 50, GC 1.03

15. AC 12, GC 1.22

17. AC 3, GC 1.6

19. AC 17, GC undefined

21. AC -15, GC 0.90

3.2 Exercises

1. 3900 customers/month

3. $337.5 per year

5. 27.05% increase

7. 6.69% decrease

9. 6.15% increase

11. 1.63% decrease

13. 6% decrease

15. Yes, you increased your sales by 113.33%

17. No, they only increased the number of students by 1.56%

3.3 Exercises

1. 2020 customers

3. 525 rabbits

5. $13 per hour

7. 25 fish

9. 81 mg

11. $16.25

13. $4.48

15. $0

17. $11.20 an hour

©Brian Gillispie, 2016.

19. 15000 rabbits

21. 153 pizzas

3.4 Exercises

1. 32 ducks

3. $4400

5. 3450 people

7. 3100 students

9. 40 ducks

11. 196.89 mg

13. 28047 bees

15. $32 - 40$ ducks

17. $10000 - 11180$ copies

19. between September 7th and September 17th

3.5 Exercises

1.

Month	Number of Pigeons	AVC
0	25	NA
3	43	6
6	67	8
9	79	4
12	82	1

3.

Hour	Cells	AVC
0	500	NA
1	550	50
2	650	100
10	2650	250
11	2675	25

5.

Year	Number of Rabbits	GC
0	12	NA
1	30	2.5
2	75	2.5
3	188	$2.50\overline{6}$
4	470	2.5

7.

Hour	Gnats	AC
0	752	NA
1	742	−10
2	532	−210
3	258	−275
4	12	−246
5	0	−12

9.

Seconds	Speed (in mph)	AC
0	5	NA
1	12	7
2	20	8
3	30	10
4	35	5
5	41	6

3.6 Exercises

1. 38 owls

3. 47 owls

©Brian Gillispie, 2016.

5. 37 goats

7. 18.8 g

9. 7 pm

11. 312 ducks

13. 5.6 liters

15. Between May 2nd and May 3rd

4.1 Exercises

1. 13

3. 53

5. 81

7. 16

9. 25

11. 607.75

4.2 Exercises

1. 12

3. 240

5. 89360

7. 530.604

9. 68

11. 1408

13. 13

15. 4255

17. 4502.096

19. 151317.8

21. 176

4.3 Exercises

1. $a_n = 0.8 * a_{n-1}, \quad a_0 = 100$

3. $a_n = 0.5 * a_{n-1}, \quad a_0 = 1000$

5. $a_n = a_{n-1} + 5, \quad a_0 = 50$

7. $a_n = a_{n-1} - 6, \quad a_0 = 40$

9. $a_n = 1.2 * a_{n-1}, \quad a_0 = 55$

11. $a_n = 0.8 * a_{n-1}, \quad a_0 = 500$

13. $a_n = 1.1 * a_{n-1}, \quad a_0 = 595$

15. $a_n = a_{n-1} + 20, \quad a_0 = 250$

17. $a_n = 1.15 * a_{n-1} - 1000, \quad a_0 = 14000$

4.4 Exercises

1. 3000, stable

3. 5000, unstable

5. 250,000, unstable

7. 2500, stable

©Brian Gillispie, 2016.

9. 5000, stable

11. $-1666.\overline{6}$, unstable

13. About 704.23 mg long term

15. No, long term there will be about 2000 rabbits left

17. Long term, there will be an infinite number of wolves

19. Yes

21. Yes

23. Long term, no people will be infected by the disease

4.5 Exercises

1. 76 rabbits, 14 wolves

3. -1.32%

5. 73 wolves, 2531 rabbits

7. 10 wolves, 70500 rabbits

9. 985 wolves, 0 rabbits

11. Bacteria A has 9 cells, Bacteria B has 14 cells

13. Store S has 920 customers, Store C has 880 customers

15. Store S has 771 customers, Store C has 1029 customers

17. 7500 susceptible, 6900 infected, 600 released

19. Long term, everyone is released

21. 3701 susceptible, 402 infected, 1897 released

©Brian Gillispie, 2016.

23. 4302 susceptible, 500 infected, 1198 released

5.1 Exercises

1. 100

3. 25

5. $y = 30x - 55$

7. $y = -6x + 25$

9. $y = -0.26x + 5.6$, x in years after 2000

11. $y = 50$

13. $y = 0.6x + 331$, x in temperature in Celsius, y in speed of sound (in mph).

15. $y = 10x + 150$, x in number of books printed, y in cost to print the books

17. $y = 6.9x - 82$, x in years after 2000, y in number of goats

5.2 Exercises

1. $y = 3x + 52$

3. $y = 2x + 15$

5. $y = 5000x + 37500$

7. 10000 copies

9. 140 deer

11. 16.8 feet

13. 54.92 degrees

15. 5.10 inches

©Brian Gillispie, 2016.

17. 30.4 degrees

5.3 Exercises

1. 49 degrees

3. $13,200

5. $64.99

7. $3.75

9. −12 Celsius

11. 45 wolves

13. 12 hours

15. July 21st

17. −8 degrees

5.4 Exercises

F1. 200 computers

F3. 12,500 copies

F5. 2,000 copies

F7. $300

F9. 365.6\overline{6}$

F11. 283.5 million dollars

S1: $y = -5000x + 60000$

S3: No

S5: $y = 8000x - 6000$

S7: Demand Equation: $y = -100x + 600$, Supply Equation: $y = 75x - 175$

S9: 300 pounds demanded, 50 pounds supplied. This is a shortage

S11: $6

S13: 5000 boxes demanded, 6250 boxes supplied. This is a surplus

S15: 3625 boxes supplied, 3500 boxes demanded. This is a surplus

6.1 Exercises

1. 78125

3. 18.52

5. 11.40

7. $x = 3$

9. $x = 3.86$

11. $x = 0.65$

13. $x = 102.10$

15. $x = 13.17$

17. $x = 62.86$

19. $x = 10$

21. $x = 8.44$

23. $x = 71.08$

©Brian Gillispie, 2016.

6.2 Exercises

1. $y = 165 * 0.98^x$

3. $y = 10 * 5^x$

5. $y = 54 * 1.1^x$

7. $y = 1400 * 1.06^x$

9. 134.82 degrees

11. 273.50 mg

13. 21.60 weeks

15. 14.81 hours

17. 2400 cells

19. 5.98 months

21. 2653.80 people

23. 690.22 catfish

6.3 Exercises

1. $y = 100 * 2.55^{(x/4)}$

3. $y = 100 * 1.5^{(x/6)}$

5. $y = 500 * 1.85^{(x/6)}$

7. $y = 50 * 0.65^{(x/20)}$

9. 16105 people

11. 186.96 mg

©Brian Gillispie, 2016.

13. 11530.54 marks

15. Around the year 2126

17. 0.39 liters

6.4 Exercises

1. $y = 150 * 2^{(x/8)}$

3. $y = 750 * 2^{(x/5.23)}$

5. $y = 91 * 2^{(x/38.75)}$

7. $y = 15 * 0.5^{(x/30.17)}$

9. 32.51 years

11. 198.24 years

13. 34.31 minutes

15. 6.84 hours

17. 21.22 years

19. 21.22 years

21. 3.72 days

23. 7.27 years

6.5 Exercises

1. $y = 50 * 1.5^{(x/8)}$

3. $y = 35 * 1.43^{(x/10)}$

5. $y = 1000 * 0.9^{(x/3)}$

©Brian Gillispie, 2016.

7. $y = 20 * 3^{((x-4)/6)}$

9. $76,081.55 per year

11. 7.38 years

13. 13.09 mSV

15. 144,000 deer

6.6 Exercises

E1. 225 rabbits

E3. 68.63 years

L1: 393 wolves

L3: 347.85 million people

L5: 71.25 minutes, or about 2:11 pm

N1: 11.73 minutes

N3: About 1:09 pm the previous day

7.1 Exercises

1. $y = -16t^2 + 32t + 80$

3. $y = -16t^2 + 250t + 2$

5. $y = -16t^2 + 1500$

7. $y = -1.33t^2 + 25.6t + 5.1$

9. 3.36 seconds

11. 9.50 seconds

©Brian Gillispie, 2016.

13. At 0.13 seconds and 2.50 seconds

15. 39.721 feet

17. No. Answers will vary for why

19. 1200 feet

7.2 Exercises

1. $y = 3.75x + 55$

3. $y = -16x^2 + 88x$

5. 2.16 seconds

7. $y = -2.5x^3 + 10.5x^2 - 10x + 8$. x in days after August 2nd

9. $y = -445x^3 + 6497.5x^2 - 23142.5x + 89747$

11. $y = -3.75x^4 + 35.8\overline{3}x^3 - 107.25x^2 + 119.1\overline{6}x + 156$, x in days after December 10th

13. 0 penguins

7.3 Exercises

1. $y = 25x$

3. $y = 2x$

5. $y = 2x^3$

7. $y = \frac{10000}{x}$

9. $y = \frac{500}{\sqrt{x}}$

11. 284.44 feet

13. $200

©Brian Gillispie, 2016.

15. 15,570.19 kilograms

17. 0.0625 watts

19. 0.216 watts

21. $7.56 * 10^{-12}$ watts

7.4 Exercises

1. Amplitude: 10, Period: 30, Phase Shift: 15, Vertical Shift: 25

3. Amplitude: 32, Period: $\frac{\pi}{2}$, Phase Shift: 13, Vertical Shift: 10

5. Amplitude: 95.2, Period: $\frac{\pi}{40}$, Phase Shift: 1.5, Vertical Shift: 4

7. Amplitude: 10, Period: $\frac{\pi}{5}$, Phase Shift: -1, Vertical Shift: 10

9. 100 birds

11. $y = 76 * \sin\left(\frac{\pi}{6}(x - 3)\right) + 80$

13. 146 ducks

15.

X	Y	Calculated Value	Absolute Error
Dec. 2009	150	152	2
June 2010	10	10	0
Dec 2010	152	152	0
June 2011	15	10	5
Dec. 2011	145	152	7
Sum			14

7.5 Exercises

1. -2, 1.18, 2.95, 5.63, 7.28

3. 0.30, 0.70, 0.89, 3.04, 8.95

©Brian Gillispie, 2016.

5, 1.20, 1.81, 5.40, 6.05, 10.04

7. 93.98, 138.45, 48.75, 86.99, 141.46

9. y = 0.574x - 0.25, x months after March 2020, 10^y is the current stock price

11. y = 0.84x + 2.3, x days after July 4th, 10^y is the current number of cells of bacteria

13. $1.58 * 10^{11}$ plants

15. 237,137.37 grams

17. 11,285 infected

19. 70.63 seconds

8.1 Exercises

1. Average: 16, Variance: 80.4, Standard Deviation: 8.97

3. Average: 16.58, Variance: 87.54, Standard Deviation: 9.36

5. Average: 252.27, Variance: 39,948.26, Standard Deviation: 199.87

7. Average: 294, Variance: 455,861.2, Standard Deviation: 675.17

9. $\mu_x = 2.5, \mu_y = 497.5, \sigma_x^2 = 1.67, \sigma_y^2 = 7275, \sigma_x = 1.29, \sigma_y = 85.29$

11. $\mu_x = 1.5, \mu_y = 0.5875, \sigma_x^2 = 1.67, \sigma_y^2 = 0.10, \sigma_x = 1.29, \sigma_y = 0.31$

13. $\mu_x = 3.28, \mu_y = 5887.5, \sigma_x^2 = 1.38, \sigma_y^2 = 256964.29, \sigma_x = 1.18, \sigma_y = 506.92$

8.2 Exercises

1. $y = 10x + 200$

3. $y = 1000 * 0.9^x$

5. $y = 15x^2 - 55x + 500$

©Brian Gillispie, 2016.

7. $3250

9. Between 2026 and 2027

11. March of next year

13. Between May 8th and 9th

15. 47,000 phones

8.3 Exercises

1. $y = 1300x + 12000$, x in months after April

3. $y = 7.25 * 1.24^{(x/2)}$, x in years after age 16

5. $y = -200x + 6000$, x in days after September 30th

7. 54 grams

9. 1,431 jars of honey

11. $24.99

13. $17.75 an hour

15. 7 million dollars

17. 15.87 million dollars

19. -14 degrees

8.4 Exercises

1. 288.75

3. 410.75

5. 8.75

©Brian Gillispie, 2016.

7. $y = 3x + 46$

9. $y = 1536 * 2.43^{(x/4)}$, x in days after January 27th

11. $y = 1300x + 12000$, x in months after April

13. $y = 7.25 * 1.24^{(x/2)}$, x in years after age 16

15. $y = -200x + 6000$, x in days after September 30th

17. 5450 people

19. $y = 22 * 1.5^{(x/3)}$, $R^2 \approx 0.77$

21. $y = 12 * 1.17^{(x/4)}$, $R^2 \approx 0.03$

23. $y = 0.2x + 4$, $R^2 \approx 0.59$

8.5 Exercises

1. 0.23

3. 0.004

5. 0.80

7. 0.85

9. Answers will vary by categorization chosen.

9.1 Exercises

1. $x = 8$

3. $x \approx 6.4061$

5. $x \approx 0.6742$

7. Answers vary, as there are three zeros and the algorithm will only find one of them.

©Brian Gillispie, 2016.

9. $x \approx 1.3797$

11. $x = -9$

9.2 Exercises

1. $y = 5x + 15, \quad 0 \leq x \leq 10$

3. $y = 0.5 * 0.55^x, \quad 0 \leq x \leq 2.69$

5. $y = 10 * 0.1^x, \quad 0 \leq x \leq 3$

7. $y = -16t^2 + 36t + 10, \quad 0 \leq x \leq 2.5$

9. $y = 50x + 450, \quad 0 \leq x \leq 11$, x in hours after 8 am

11. $y = -75x^2 + 8250x, \quad 22.98 \leq x \leq 87.02$

13. $y = -5x^2 + 70x + 100, \quad 3 \leq x \leq 10$, x set up such that $x = 3$ means March, and so on

9.3 Exercises

1. -6

3. 127.81

5. 200

7. 144

9. 239

11. $f(x) = \begin{cases} 400 * 1.05^x, & 0 \leq x < 14.21 \\ 800, & 14.21 \leq x \end{cases}$

13. $f(x) = \begin{cases} -16t^2 + 300t + 2, & 0 \leq x < 18.78 \\ -16(t-18.78)^2 + 60(t-18.78), & 18.78 \leq x < 20.72 \\ 0, & 20.72 \leq x \end{cases}$

17. After the 4th raise

©Brian Gillispie, 2016.

19. 4.36 years

21. $2.11 per pound

23. Roughly the years 2024 and 2027

9.4 Exercises

1. $x = 14, y = 71.64$

3. $x = 475.2, y = 332.56$

5. 3.96 feet

7. 125.35 feet

9. -105.38 feet. Reasons for the answer will vary

11. 2819.53 feet

13. No, the ball does not clear the fence

15. Yes, the kick clears the crossbar

17. Yes, the kick clears the crossbar

Index

AC. See arithmetic change
amplitude, 386
ARC. See average rate of change
arithmetic change
    definition of, 128
    in recursive sequence, 202
    interpreting, 135
    use in change comparison, 428
average. See mean
average rate of change
    between two data points, 126
    calculation of, 126
    definition of, 125
    interpreting, 134
    using as slope, 248
    using for predictions, 142, 153, 159
bisection method. See zero finding
break-even point, 266
categorical data
    converting to quantitative, 95
change
    arithmetic change, 128
    average rate of change, 125
    combined change, 148
    geometric change, 129
    total change, 123
change of the arithmetic change
    calculation of, 171
    definition of, 170
    method of change comparison, 428
    notation of, 170
    predicting with, 173
change table, 164
cost equation, 264
demand equation, 269
domain
    definition of, 100
    of exponential, 105
    of line, 105
    of logarithmic, 107
    of quadratic, 106
    of trigonometric, 106
    restricting of, 482
doubling time, 309
e, 285
equation
    area, 54
    distance-rate-time, 58
    evaluating, 21
    perimeter, 54
error
    definition of, 39
    random, 43
    rounding, 44
    systematic, 43
    types of, 43
error tables
    how to make, 46
    uses in modeling, 49, 438, 447
exponential model
    continuous growth model, 328
    domain of, 105
    doubling time, 310
    finding doubling time, 313
    finding half life, 314
    from geometric change, 292
    from percent decrease, 292
    from percent increase, 292
    from two data points, 323
    half life, 310
    logistic model, 331
    method of change comparison, 428
    method of least error, 438
    method of lowest squared error, 447
    newton's law of cooling, 336
    range of, 105
    time units of x, 295
exponents
    definition of, 283
    rules of, 284
fixed cost, 263
fixed point

definition of, 210
stability of, 214
function
    piecewise, 489
GC. See geometric change
geometric change
    definition of, 129
    in recursive sequence, 202
    interpreting, 136
    use in change comparison, 428
    using for predictions, 145, 157, 159
half life, 309
height equation
    equation of, 347
    initial height, 347
    initial velocity, 347
    parametric version of, 504
heizenburg uncertainty principle, 114
institutional review board, 116
inverse square law. See proportionality
line. See linear model
linear model
    break-even point, 266
    cost equation, 264
    demand equation, 269
    domain of, 105
    from two data points, *256*
    method of change comparison, ***428***
    method of least error, 438
    method of lowest squared error, 447
    point-slope form, *243*
    range of, 105
    slope intercept form, *241*
    slope of, *237*
    supply equation, 270
    y-intercept, *240*
literary digest example, 15
log scaling
    creating a scaled model, 401
    idea of, 397
logarithm
    definition of, 285
    rule of, 286
logarithmic function
    domain and range of, 106
logistic model, 331

mean, 413
method of change comparison, 428
method of least error, 438
method of lowest squared error, 447
model creation, 15
model selection
    amount of variance left in model, 449
    method of change comparison, 428
    method of least error, 438
    method of lowest squared error, 447
modeling
    categorical data, 458
    categorical to quantitative, 95
    cautions on, 18
    changing scales, 89
    domain of, 100
    domain restricting, **482**
    ethical issues, data collection, 113
    ethical issues, data reporting, 115
    height equation, 347
    overflow, 24, 25
    piecewise model, 491
    polynomial model, 361
    range of, 104
    scientific notation, 23
    setting up data points, 80
    time, 79
    trigonometric model, 391
    underflow, 24, 25
    unlike units, 26
    using change, 153, 157, 173
    using scaled model, 85
    variable setup, 80
newton's law of cooling, 336
number scaling
    log scaling, 401
observational study, 114
parametric model
    definition of, 501
    evaluation of, 501
    height equations, 504
percent decrease
    from geometric change, 136
    to geometric change, 147
percent increase
    from geometric change, 136

©Brian Gillispie, 2016.

to geometric change, 146
period, 385
phase shift, 383
piecewise function
   creation of, 491
   definition of, 489
   evaluating, 489
polynomial interpolation. See polynomial model
polynomial model, 361
   issues with, 366
predator-prey model, 224
proportionality
   direct, 371
   inverse, 374
   inverse square law, 376
quadratic. See quadratic model
quadratic model
   domain of, 106
   maximum of, 354
   method of change comparison, 428
   minimum of, 354
   range of, 106
   zeros of, 348
random error, 43
range
   definition of, 104
   of exponential, 105
   of line, 105
   of logarithmic, 107
   of quadratic, 106
   of trigonometric, 106
recursive sequence
   fixed point of, 210
   interconnected sequence, 224
   representation of, 192
   using rates of change, 204
recursive sequences
   from two points, 201
   in excel, 197
relative error
   definition of, 40
revenue equation, 265
rounding error, 44
scaled model
   use of, 85

secant method. See zero finding
sequence
   definition of, 187
   notation, 188
   recursive notation, 192
shortage, 273
S-I-R model, 228
slope, 237
SSE. See sum of squared errors
standard deviation, 416
sum of squared errors, 447
supply equation, 270
surplus, 273
systematic error, 43
TC. See total change
total change, 123
   issues with, 124
trigonometric model
   amplitude, 386
   doman of, 106
   error of, 394
   how to create, 391
   issues with, 394
   period, 385
   phase shift, 383
   range of, *106*
   vertical shift, 387
units
   adding, 29
   dividing, 33
   multiplying, 31
   subtracting, 29
variable cost, 263
variables
   categorical, 69
   continuous, 73
   dependent, 77
   discrete, 73
   from table, 82
   independent, 77
   interval, 71
   levels of measurement, 71
   nominal, 71
   ordinal, 71
   quantitative, 69
   ratio, 72

time, 79
types of, 69
variance, 421
vertical shift, 387
vertical translation. See vertical shift

y-intercept, 240
zero finding
    bisection method, 473
    secant method, 478

References

[1] M. Levy. [Online]. Available: https://www.britannica.com/event/United-States-presidential-election-of-1936. [Accessed 8 December 2016].

[2] D. Leip. [Online]. Available: http://uselectionatlas.org/RESULTS/national.php?year=1936&f=0. [Accessed 8 December 2016].

[3] D. S. Moore, The Basic Practice of Statistics, 5th edition, New York: W.H Freeman and Company, 2010.

[4] "Gallop Student Poll," 2013. [Online]. Available: http://www.gallupstudentpoll.com/home.aspx. [Accessed 4 January 2014].

[5] F. McCown, "Schelling's Model of Segregation," [Online]. Available: http://nifty.stanford.edu/2014/mccown-schelling-model-segregation/. [Accessed 8 December 2016].

[6] P. G. Zimbardo, "Stanford Prison Experiment," Social Psychology Network, [Online]. Available: http://www.prisonexp.org/. [Accessed 8 December 2016].

[7] S. Lynch, Dynamical Systems with Applications, using Maple, Boston: Birkhauser, 2001.

[8] F. R. Marotto, Introduction to Mathematical Models Using Discrete Dynamical Sytems, Belmont: Thomson Brooks/Cole, 2006.

[9] R. McCormick, "A North Dakota town is the most expensive place to rent an apartment in the United States," The Verge, 19 February 2014. [Online]. Available: http://www.theverge.com/2014/2/19/5425040/williston-north-dakota-most-expensive-place-to-rent-in-us. [Accessed 8 December 2016].

[10] J. Healy, "Built Up by Oil Boom, North Daktoa Now Has an Emptier Feeling," New York Times, 7 February 2016. [Online]. Available: http://www.nytimes.com/2016/02/08/us/built-up-by-oil-boom-north-dakota-now-has-an-emptier-feeling.html?_r=0. [Accessed 8

©Brian Gillispie, 2016.

December 2016].

[11] G. Oster, "Thomas Austin and His Rascally Rabbits," Hankering for History, 18 January 2014. [Online]. Available: http://hankeringforhistory.com/thomas-austin-and-his-rascally-rabbits/. [Accessed 8 December 2016].

[12] C. a. M. Kime, Explorations in College Algebra, 5th edition, John Wiley and Sons, Inc., 2011.

[13] K. E. Atkinson, An Introduction to Numerical Analysis, Wiley, 1989, p. 712.

[14] A. Argesti, "Applying R(squared) -Type Measures to Ordered Categorical Data," *Technometrics*, pp. 133 - 138, 1986.

[15] M. Celeste Robb-Nicholson, "A doctor talks about: Radiation risk from medical imaging," Harvard Medical School, October 2010. [Online]. Available: http://www.health.harvard.edu/cancer/radiation-risk-from-medical-imaging. [Accessed 8 December 2016].

©Brian Gillispie, 2016.

Made in the USA
San Bernardino, CA
12 January 2017